Hubble 2008
Science Year in Review

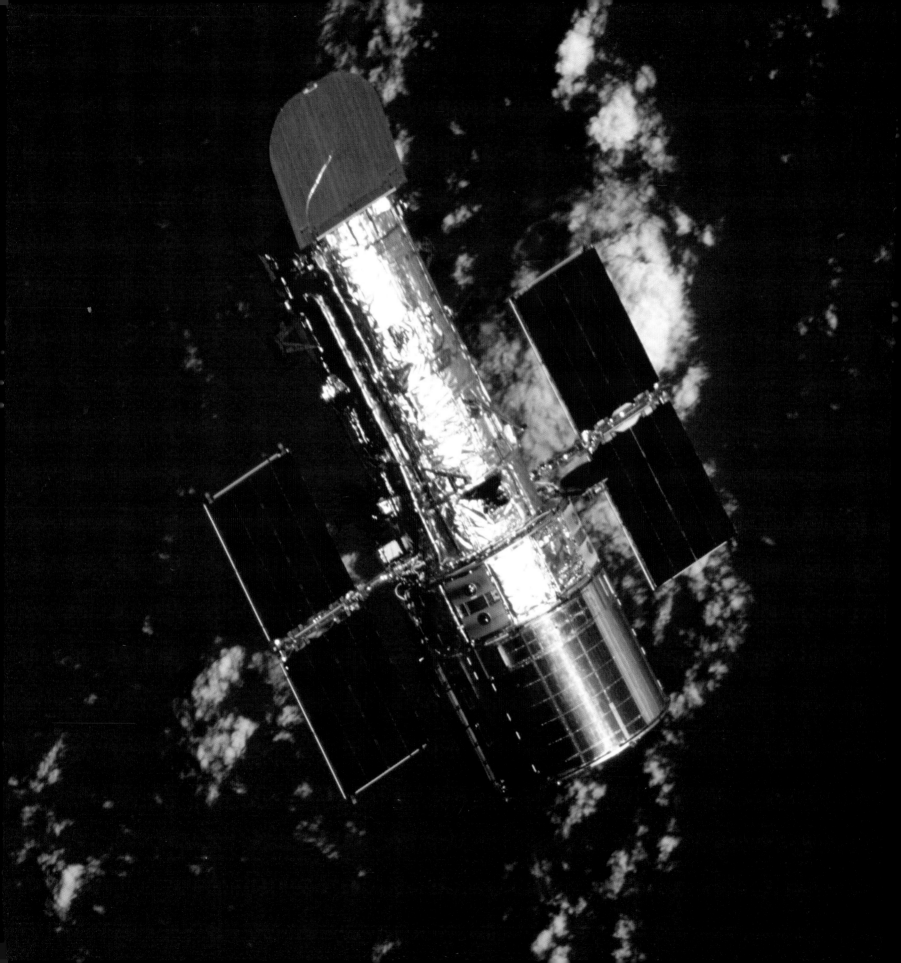

Foreword	5
Hubble's History	9
Observatory Design	19
Operating *Hubble*	25
Servicing Mission 4 Preparation	35
Hubble News	45
Science	**53**
A New Red Spot on Jupiter	55
Dwarf Bodies in the Solar System	61
First Visible-Light Image of an Extrasolar Planet	73
Supernova Remnant SN 1006	83
Probing the Atmospheres of Exoplanets	89
Magnetic Filaments in an Active Galaxy	101
Interacting Galaxies	105
Dark Matter and Galaxy Life in a Supercluster	113
Barred Spiral Galaxies and Galactic Evolution	121
Searching for Baryonic Matter in Intergalactic Space	125
Supporting *Hubble*: Profiles	**133**
Acknowledgments	**149**

Hubble Space Telescope as seen from the Space Shuttle *Columbia* following its March 2002 servicing call. On-orbit upgrades have kept the telescope on the cutting edge of astronomical research.

Foreword

Hubble Space Telescope is the flagship of the NASA great observatories. When it was lofted into Earth orbit in 1990, *Hubble* quickly and dramatically launched a golden age of space astronomy that has simply been unprecedented in the history of science. *Hubble*'s never-before-seen, razor-sharp views of the near and far universe have charted a previously "undiscovered country" with an intimacy, clarity, and new depth of knowledge not realized since Galileo first made a telescopic survey of the heavens.

The cosmos was certainly a more staid-looking place before *Hubble*'s tenfold increase in optical resolution unveiled a cataclysmic, discordant, and evolving universe where matter and energy shape stars, planets, and nebulas against a backdrop of galaxies that seems unimaginably deep.

Every new image reminds us what an astonishing new chapter of astronomical understanding—for scientists and laypeople alike—the *Hubble Space Telescope* has opened. *Hubble* has unveiled secrets of the universe that humans had only been able to probe in their imaginations through much of recorded history. *Hubble* has carried us on a space odyssey of discovery to distant places and times unreachable by physical travel across space.

Now, nearly 20 years after launch, the scale of the *Hubble* revolution is becoming increasingly apparent: black holes are common to galaxies; planetary atmospheres have organic chemistry; dark energy seems to behave like Einstein's cosmological constant; galaxies were rapidly born and quickly evolved through collisions and mergers; and stars die in a blaze of glory that is as awesome as it is foreboding.

Hubble has grown enormously in capability and outlived its projected operational life of 15 years, thanks to orbital servicing by Space Shuttle astronauts. With each new servicing mission, the telescope's scientific strength has been significantly increased, giving *Hubble* a boost in observing power over other space science platforms that are not serviceable. This has consistently expanded the "discovery space" available to astronomers to do cutting-edge research.

Hubble is a trailblazer for future space observatories, including the *James Webb Space Telescope* and a possible dedicated mission to study the mysterious dark energy. Astronomers are also hard at work developing design concepts for a possible future "daughter of *Hubble*," a *Hubble*-like telescope for observations of ultraviolet, optical, and near-infrared light, but with a much larger aperture of 9 to 16 meters. *Hubble* is bequeathing to the future both an amazing legacy of scientific discoveries and the impetus to continue humankind's quest to understand the universe in which we live.

Bright knots of glowing gas light up the spiral arms in nearby galaxy M74, indicating rich regions of star formation. M74 is located roughly 32 million light-years away in the direction of the constellation Pisces, the Fish.

"The history of astronomy is a history of receding horizons."

Edwin Hubble

Galaxy NGC 300 is located 7 million light-years distant in the direction of the constellation Sculptor. This image is a composite of three separate exposures taken through red, green, and blue filters by *Hubble*'s Advanced Camera for Surveys. It is part of the ANGST Galaxy Survey described in the Operating *Hubble* section of this book.

Hubble's History

Hubble's remarkable mission has now spanned 18 years. During that time, it has been at the nexus of perhaps the most exciting period of discovery in the history of astronomy. Simultaneously *Hubble* has offered up some of the most daunting engineering challenges to humans working in space, and success in meeting those challenges has been among NASA's greatest triumphs.

Since its launch in 1990, *Hubble* has been visited four times by astronauts to fix, restore, and upgrade its equipment. In nearly constant use between these servicing missions, *Hubble* has generated data for thousands of scientific papers on topics ranging from nearby planets to distant galaxies, and from forming solar systems to devouring black holes.

The concept of a large telescope in space is as old as the space program itself. In a classified study in 1946, Lyman Spitzer first articulated the scientific and technical rationale for space astronomy. He continued to be the champion of the dream of a large telescope in space until it was realized. Supported by colleagues John Bahcall, George Field, and others, Spitzer was a tireless advocate within the astronomical community, to the public, and to the Federal Government. The outcome was a "new start" for the mission, authorized by Congress in 1977.

The technology needed for the *Hubble Space Telescope* was well advanced when work began. However, other serious technological and management challenges characterized the tumultuous years of *Hubble*'s design and manufacture. This turmoil culminated with the tragic loss of Space Shuttle *Challenger* and its crew in January 1986. Finally, against the backdrop of unrestrained anticipation by the public and the astronomical community alike, NASA launched *Hubble* into orbit on Space Shuttle *Discovery* (STS-31) on April 24, 1990.

Hubble is seen here being assembled in the clean room of the Lockheed Missile and Space Company during the 1980s. The Optical Telescope Assembly, which contains the primary and secondary mirrors, is being hoisted atop the aft shroud. Yellow handrails seen on the spacecraft's body will allow astronauts to move about and service the telescope.

December 8, 1993. Astronaut Kathryn Thornton maneuvers the Corrrective Optics Space Telescope Axial Replacement unit toward the aft shroud of *Hubble* while fellow astronaut Thomas Akers guides her approach from inside.

Hubble's first few months were disastrous. Instead of returning crisp, point-like images of stars, its images showed stars surrounded by large, fuzzy halos of light. The source of the problem was traced to an error in constructing the equipment used to test *Hubble*'s mirror during manufacture. Optical tests using this equipment led technicians to grind the mirror to the wrong shape, giving it a classic case of "spherical aberration." The mirror was perfectly smooth, but it would not focus light to a single point.

There was, however, an opportunity to fix *Hubble*. The telescope was designed so astronauts could periodically upgrade and service it on orbit. Even before launch, NASA had begun to build a second-generation camera to replace the main camera that was launched with the telescope. Optical experts realized they could build corrective optics into the camera to counteract the flaw in the *Hubble* mirror. NASA accelerated work on the Wide Field Planetary Camera 2 (WFPC2), and *Hubble* scientists and engineers designed a mechanical fixture called Corrective Optics Space Telescope Axial Replacement (COSTAR) to deploy corrective optics in the light paths to the other instruments. In December 1993, astronauts returned to *Hubble* and undertook an ambitious set of space walks to install the new equipment. The modifications worked flawlessly, restoring *Hubble*'s image quality to original expectations.

These two images of the spiral galaxy M100 show the dramatic improvement to *Hubble*'s focus accomplished during the first Servicing Mission (SM1, December 1993). Spherical aberration in the optical figure of the telescope's primary mirror was overcome by the introduction of carefully designed corrective optics. The mission patch for SM1 appears along with these before and after shots.

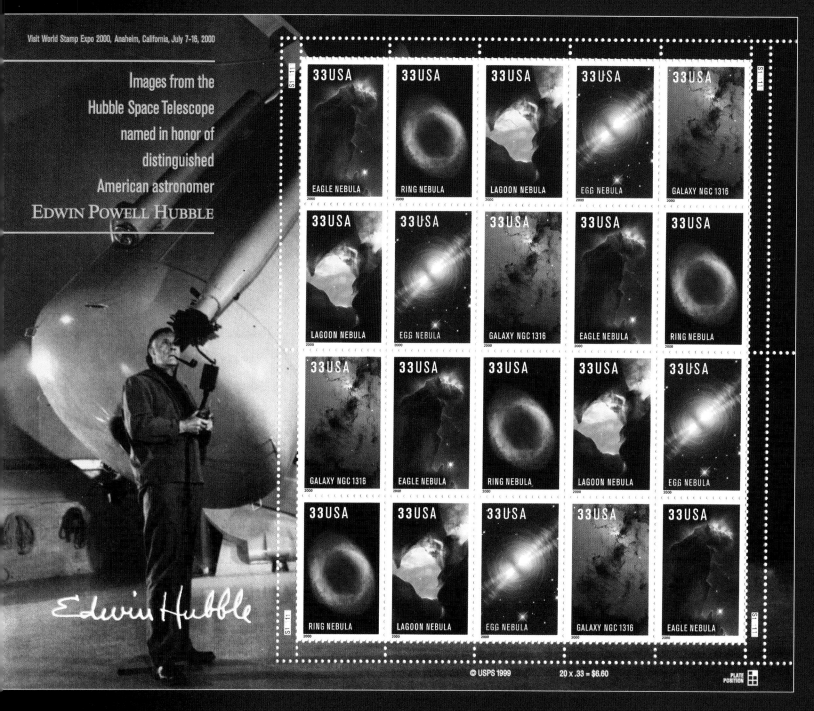

Edwin Hubble, for whom the *Hubble Space Telescope* is named, was one of the leading astronomers of the twentieth century. His discovery in the 1920s that countless galaxies exist beyond our own Milky Way galaxy revolutionized our understanding of the universe. Hubble's perhaps most notable contribution, however, was his observation that the farther apart galaxies are from each other, the faster they move away from each other. Based on this discovery, Hubble concluded that the universe expands uniformly. Several scientists had also posed this theory based on Einstein's General Relativity, but Hubble's data, published in 1929, helped convince the scientific community. In 2000, the United States Postal Service commemorated Hubble and his namesake—the *Hubble Space Telescope*—with a commemorative issue of stamps.

In the 15 years following the first servicing mission, *Hubble* has treated astronomers and the public to the clearest and deepest views of the universe—scenes of profound beauty and intellectual challenge. Thousands of astronomers have used *Hubble* for boundary-breaking research in virtually all areas, from our own solar system to the farthest depths of the expanding universe. Three additional servicing missions in 1997, 1999, and 2002 punctuated this era, and a final mission to upgrade and refurbish *Hubble* is planned for 2009.

The 1997 mission brought tremendous improvements to *Hubble*'s spectroscopic capabilities with the insertion of the Space Telescope Imaging Spectrograph (STIS). STIS observations not only demonstrated that black holes are ubiquitous in the centers of galaxies, but also showed that the black hole masses are tightly correlated with the masses of the surrounding ancient stellar population. The 1997 mission also opened *Hubble*'s view to the near-infrared universe with the Near-Infrared Camera and Multi-Object Spectrometer (NICMOS). The clear views of distant galaxies provided by NICMOS have supplied a wealth of clues to the complex physics in the early universe that led to the formation of the Milky Way.

The servicing mission in 1999 enhanced many of *Hubble*'s subsystems, including the central computer, a new solid-state data-recording system to replace the aging magnetic tape drives, and the gyroscopes needed for pointing control. A month prior to launch, a gyroscope failure had forced *Hubble* into "safe mode," with no ability to observe astronomical targets.

When a premature loss of solid-nitrogen coolant cut short NICMOS's operational life, NASA engineers used innovative mechanical refrigeration technology to develop an alternate way of cooling its detectors to their operating temperature of −320°F. This cooling system was installed in 2002, and it brought the instrument back to life. NICMOS has proved crucial to observations of very distant supernovas used to measure the acceleration of the universe. The 2002 mission also introduced *Hubble*'s most powerful camera—the Advanced Camera for Surveys (ACS)—providing a tenfold improvement over WFPC2.

In the final servicing mission in 2009, astronauts will install two new instruments, the Cosmic Origins Spectrograph (COS) and Wide Field Camera 3 (WFC3). COS is the most sensitive ultraviolet spectrograph ever built for *Hubble*. The instrument will probe the cosmic web—the large-scale structure of the universe—whose form is determined by the gravity of dark matter and is traced by the spatial distribution of galaxies and intergalactic gas. WFC3 is a new camera that is sensitive across a wide range of wavelengths (colors), including infrared, visible, and ultraviolet light. It will study planets in our solar system, the formation histories of nearby galaxies, and early and distant galaxies beyond *Hubble*'s current reach.

Attempts will also be made to repair two instruments currently installed in *Hubble*: STIS and ACS. STIS was installed in 1997 and stopped working in 2004. When repaired, the instrument will be used for high-resolution studies in visible and ultraviolet light of both nearby star systems and distant galaxies, providing information about the motions and chemical makeup of stars, planetary atmospheres, and other galaxies. ACS suffered a partial failure in early 2007 after operating exquisitely for nearly five years. Astronomers hope that it can be restored to its full capability to perform high-efficiency imaging in both the visible and ultraviolet portions of the electromagnetic spectrum.

Spacewalking astronauts will also install a refurbished Fine Guidance Sensor to replace one degrading unit of the three already onboard. Two of these sensors are routinely used to enable *Hubble*'s precise pointing, and the third is available to astronomers for making accurate measurements of stellar positions. Astronauts will also exchange a command and data handling unit that stores and transmits science data to Earth. In addition, they will replace all six of the telescope's batteries and all six gyroscopes, add new thermal coverings, and install a soft-capture mechanism on *Hubble*'s aft bulkhead.

The *Hubble Space Telescope*, operating at the intersection of the robotic and the human space flight programs, embodies both the trials and triumphs of the space program. It has survived controversy, delays, and failures, and has proven to be one of the most powerful and productive scientific tools ever developed.

In a dramatic nighttime launch, the Space Shuttle *Columbia* lifts off to commence Servicing Mission 3B on March 1, 2002.

The Hubble Space Telescope

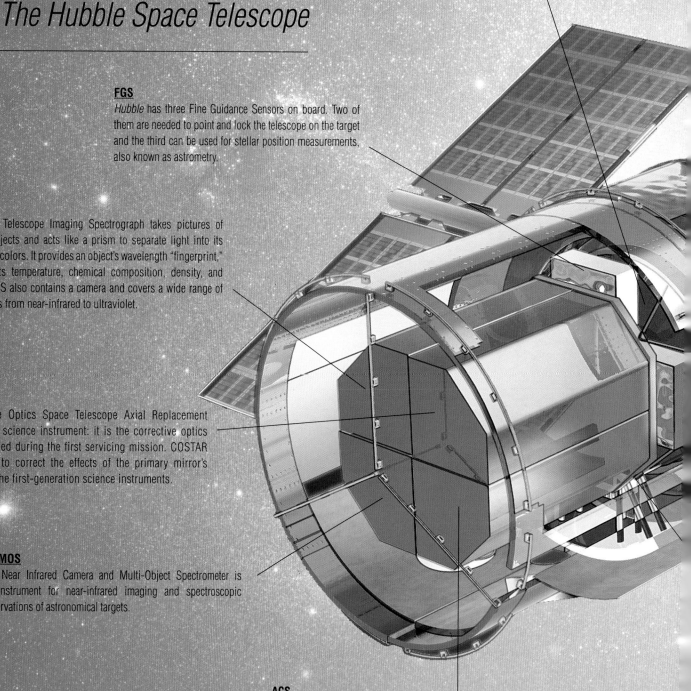

Primary mirror
Hubble's primary mirror is made of a special glass coated with aluminum and a compound that reflects ultraviolet light. It is 2.4 m in diameter and collects the light from stars and galaxies and reflects it to the secondary mirror.

FGS
Hubble has three Fine Guidance Sensors on board. Two of them are needed to point and lock the telescope on the target and the third can be used for stellar position measurements, also known as astrometry.

STIS
The Space Telescope Imaging Spectrograph takes pictures of celestial objects and acts like a prism to separate light into its component colors. It provides an object's wavelength "fingerprint," including its temperature, chemical composition, density, and motion. STIS also contains a camera and covers a wide range of wavelengths from near-infrared to ultraviolet.

COSTAR
The Corrective Optics Space Telescope Axial Replacement is not really a science instrument: it is the corrective optics package installed during the first servicing mission. COSTAR was designed to correct the effects of the primary mirror's aberration on the first-generation science instruments.

NICMOS
The Near Infrared Camera and Multi-Object Spectrometer is an instrument for near-infrared imaging and spectroscopic observations of astronomical targets.

ACS
The Advanced Camera for Surveys is designed primarily for wide-field imagery in the visible wavelengths, though it also sees in ultraviolet and near infrared. It has a wide-field channel for efficient surveys of the universe, a high-resolution channel for investigations needing the sharpest images, and a channel for observations at far-ultraviolet wavelengths.

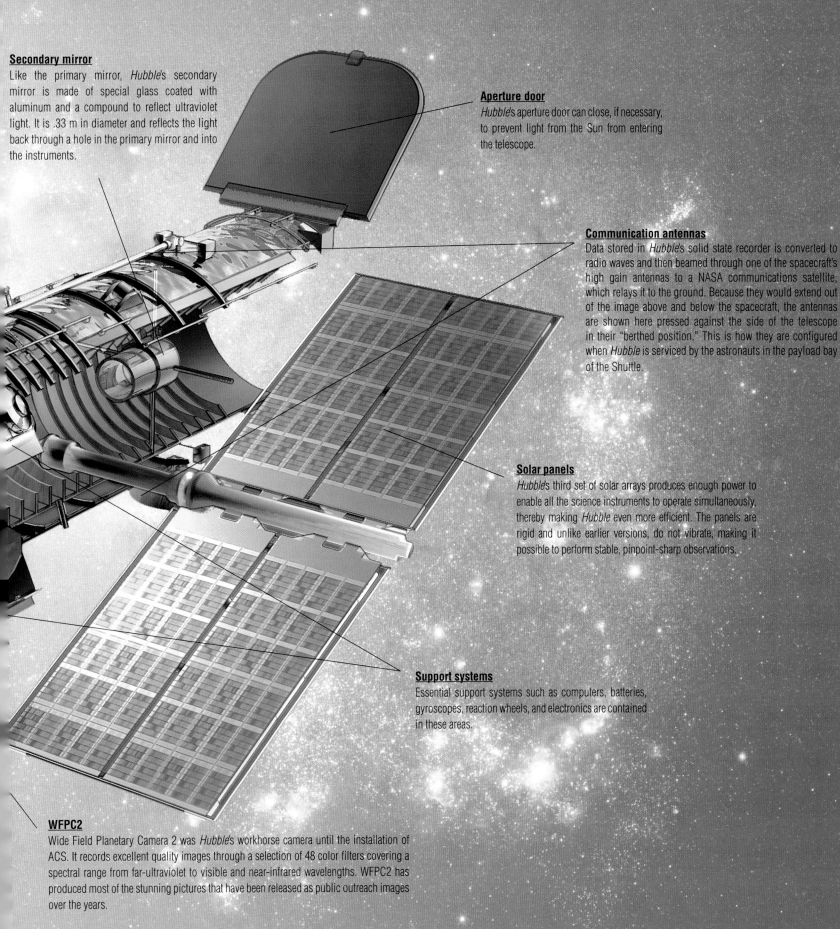

Secondary mirror
Like the primary mirror, *Hubble*'s secondary mirror is made of special glass coated with aluminum and a compound to reflect ultraviolet light. It is .33 m in diameter and reflects the light back through a hole in the primary mirror and into the instruments.

Aperture door
Hubble's aperture door can close, if necessary, to prevent light from the Sun from entering the telescope.

Communication antennas
Data stored in *Hubble*'s solid state recorder is converted to radio waves and then beamed through one of the spacecraft's high gain antennas to a NASA communications satellite, which relays it to the ground. Because they would extend out of the image above and below the spacecraft, the antennas are shown here pressed against the side of the telescope in their "berthed position." This is how they are configured when *Hubble* is serviced by the astronauts in the payload bay of the Shuttle.

Solar panels
Hubble's third set of solar arrays produces enough power to enable all the science instruments to operate simultaneously, thereby making *Hubble* even more efficient. The panels are rigid and unlike earlier versions, do not vibrate, making it possible to perform stable, pinpoint-sharp observations.

Support systems
Essential support systems such as computers, batteries, gyroscopes, reaction wheels, and electronics are contained in these areas.

WFPC2
Wide Field Planetary Camera 2 was *Hubble*'s workhorse camera until the installation of ACS. It records excellent quality images through a selection of 48 color filters covering a spectral range from far-ultraviolet to visible and near-infrared wavelengths. WFPC2 has produced most of the stunning pictures that have been released as public outreach images over the years.

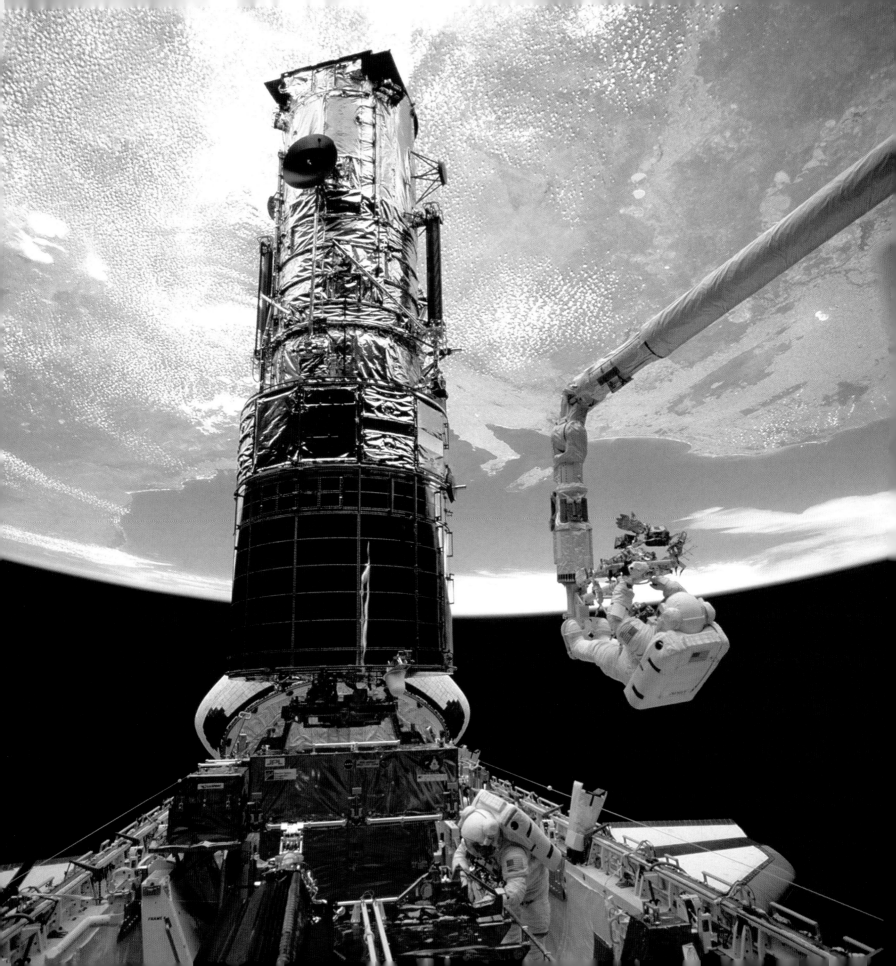

Observatory Design

With its modular instruments, handrails, and latches, *Hubble* was designed for astronaut-friendly on-orbit servicing. About the size and weight of a subway car, it filled the entire payload bay of the Space Shuttle when carried to orbit in 1990.

The heart of *Hubble* is its 2.4-m mirror, which collects about 40,000 times as much light as the human eye. High above the distorting effects of Earth's atmosphere, *Hubble* obtains images with 10 times the typical sharpness of ground-based telescopes and views wavelengths of near-infrared and ultraviolet light that do not reach the Earth's surface.

Hubble has an optical layout known as a Ritchie-Chrétien Cassegrain design. The incoming light bounces off the primary mirror, up to a secondary mirror, and back down through a hole in the primary mirror, where it comes to a focal plane that is shared among the suite of scientific instruments. A graphite-epoxy truss provides a rigid structure for the main optics and a system of baffles painted flat black is mounted within the telescope to suppress stray or scattered light from the Sun, Moon, or Earth.

Hubble is encased in a thin aluminum shell, blanketed by many thin layers of insulation, which protects against the widely varying temperature fluctuations experienced on orbit. The telescope itself is housed in the narrower top section of the tube. Most of the control electronics sit in the middle of the telescope, where the tube widens. The middle section also houses *Hubble*'s four 100-pound reaction wheels. *Hubble* reorients itself around the sky by exchanging momentum with these spinning flywheels. Astronauts can easily access the devices in *Hubble*'s midsection, and a number of these have been replaced or upgraded during servicing missions. At the back end of the spacecraft, the "aft shroud" houses the scientific instruments, gyroscopes, star trackers, and other components.

> While Astronaut Jeffrey Hoffman works in the payload bay, Astronaut F. Story Musgrave—anchored on the Space Shuttle *Endeavour*'s robotic arm—prepares to be elevated to the top of *Hubble* during Servicing Mission 1 (SM1, December 1993). During SM1, the observatory was refurbished and improved in a number of significant ways, as it has during each of the subsequent Servicing Missions (SM2, SM3A, and SM3B). All the improvements were enabled by the telescope's forward-thinking, modular design.

All of the spacecraft's interlocking shells—the light shield, forward shell, equipment section, and aft shroud—provide a benign thermal and physical environment in which sensitive telescope optics and scientific instruments can operate properly for many years. Excluding the aperture door and solar arrays, *Hubble* is about 43 ft in length and 14 ft in diameter at its widest point. Altogether, it weighs about 25,000 pounds.

Hubble's electrical power comes from two 25-ft long solar panels, which are mounted like wings on the side of the observatory and rotate to point toward the Sun. Six batteries, charged by solar power, provide power when the telescope is in shadow. Astronauts replaced the solar arrays on two occasions during servicing missions. The present arrays are rigid panels of gallium arsenide cells, which were originally designed for commercial communications satellites. They are about 30% more efficient in converting sunlight to electricity than the prior arrays. When new, they generated about 5,700 W of electrical power.

In a single orbit around Earth, the exterior surface of *Hubble* varies in temperature from −150°F to +200°F. Despite the harsh thermal environment, the interior of *Hubble* is maintained within a narrow range of temperatures—in many areas at a "comfortable room temperature"—by its thermal control system. Temperature sensors, electric heaters, radiators, insulation inside the spacecraft and on its outer surface, and paints that have special thermal properties, all work in concert to minimize the expansion and contraction that could throw the telescope out of focus, and to keep the equipment inside the spacecraft at proper operating temperatures. In addition to guiding the telescope, the fine guidance sensors are used to make very precise measurements of the relative positions of stars, which are essential for estimating distances to nearby stars or masses of components of binary star systems.

The aft shroud has room for five scientific instruments. Over the years, NASA and the European Space Agency (ESA) have manufactured 14 scientific instruments for *Hubble*. Each new generation of instruments has brought enormous improvements to the scientific capabilities of the observatory through advances in technology. Many of *Hubble*'s discoveries with these new instruments would have been impossible to achieve with the instruments installed at launch.

The Cosmic Origins Spectrograph resides in its protective enclosure as engineers and technicians prepare it for Servicing Mission 4. It is designed for sensitive ultraviolet spectroscopy of faint point sources. Science goals for the instrument include the study of the origins of large scale structure in the universe, the formation and evolution of galaxies, and the origin of stellar and planetary systems and the cold interstellar medium.

Hubble uses six nickel hydrogen batteries—two modules of three batteries each—to store electricity and power its systems. Combined, they deliver over 450 amp-hours of charge to the spacecraft. Through careful management, *Hubble*'s original six batteries have lasted more than 13 years longer than their design life of 5 years—but are now scheduled for replacement during Servicing Mission 4 (SM4). Each battery contains 22 individual cells wired together in series, and an isolation switch that electrically disables the connectors, making them safe for handling by the astronauts. A module containing three batteries (seen in the top photo with its top cover removed) weighs 460 pounds. Mission Specialist Michael Good is seen in the bottom photo, practicing the installation of one of these modules into an SM4 mechanical simulator.

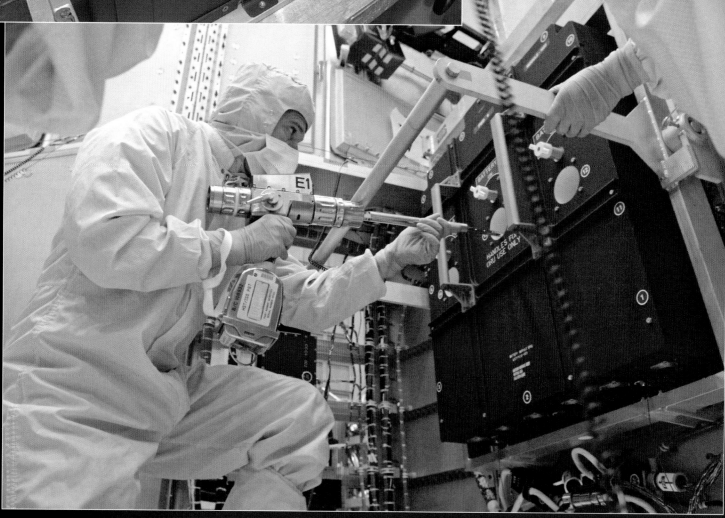

At Kennedy Space Center, Wide Field Camera 3 is placed in its protective enclosure in preparation for launch on SM4. The instrument was designed as a versatile camera capable of imaging astronomical targets over a very wide wavelength range and with a large field of view. It is a fourth-generation instrument, making use of structural components from the original camera. Wide Field Camera 3 has two independent light paths: an optical channel, and a near-infrared channel.

On Saturday, September 27, 2008 shortly after 8 p.m., *Hubble* experienced a failure of the Side A Control Unit/Scientific Data Formatter (CU/SDF), which is responsible for storing, formatting, and transmitting *Hubble* science data to the ground. The CU/SDF is a part of the Scientific Instrument Command and Data Handling (SIC&DH) unit (pictured here) that resides on the Bay 10 door of *Hubble*. The operations team was able to transition to Side B of the CU/SDF, returning the telescope to scientific operation. However, the importance of maintaining redundancy in this unit forced a delay of SM4, which was scheduled to launch just a few weeks later. NASA now plans to replace the entire SIC&DH during SM4, which was delayed sufficiently to allow for testing and qualification of the replacement unit.

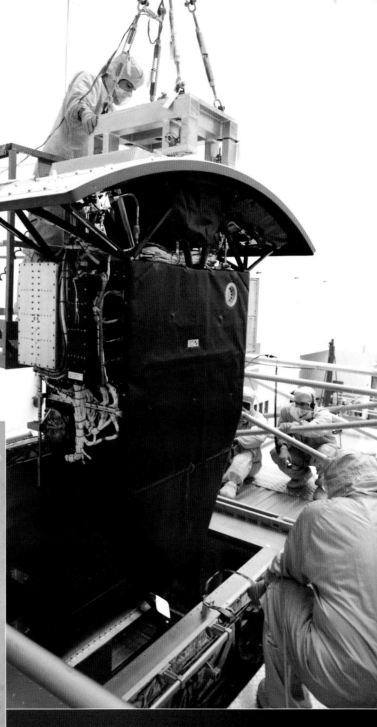

Operating *Hubble*

Circling Earth at an altitude of 360 miles, *Hubble* completes an orbit every 96 minutes. This constant movement from light to shadow requires science operations to be carefully planned to optimize the observing schedule. Planning is made even more complex because targets may be visible at only certain times of the year. The carefully assembled calendar intersperses observations from different programs to maximize the telescope's observing efficiency.

It is the job of *Hubble* controllers at the Space Telescope Science Institute and NASA's Goddard Space Flight Center to seamlessly blend science operations and spacecraft operations 24 hours a day. Scientists and engineers at the Institute translate the research plans of astronomers into detailed sequences of commands for the internal electronics, detectors, and mechanisms of the scientific instruments. The preparations, carried out weeks or months in advance of the observations, also involve selecting guide stars to stabilize the telescope pointing.

Spacecraft controllers work together to schedule *Hubble*'s communication with the ground, to load commands into the onboard computers, to configure the distribution of electrical power from solar arrays and batteries, and to manage the data in the onboard computers. The flight operations team at Goddard monitors every system on *Hubble* to ensure it is working properly. If one is not, ground controllers can intervene to remedy the problem—if the onboard safing system has not already done so autonomously.

Calls for proposals to use *Hubble* are issued annually, and anyone can apply for observing time. The application process is open to worldwide competition without restrictions on nationality or academic affiliation. Potential users must show that the observations can only be accomplished with *Hubble*'s unique capabilities and are beyond the capacity of ground-based telescopes.

A team of dedicated, detail-oriented individuals work together to keep *Hubble* safe and productive. Both the spacecraft, and the ground-elements of the overall system need to be configured, monitored, and maintained.

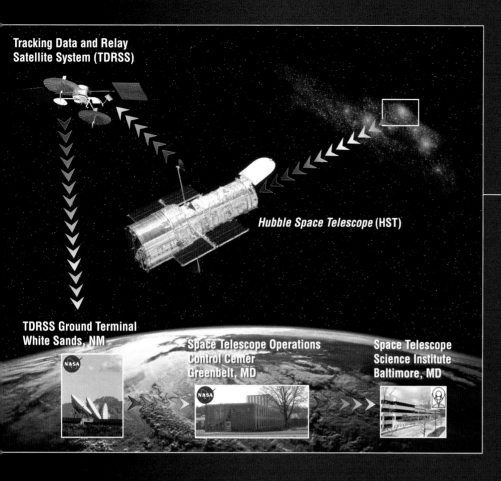

Hubble data is transmitted to Earth through a NASA relay satellite, which downlinks it to a ground station in White Sands, NM. From there, it is forwarded to Goddard Space Flight Center for initial processing and quality checking. Within minutes, it is sent to the Space Telescope Science Institute, where it is further processed, archived, and made available to the Principal Investigator who successfully proposed the observation.

Competition for time on the telescope is extremely intense. The demand is so great that there are typically six times more proposals to use the telescope than are actually selected. Telescope observing time is measured by the number of orbits required for a successful observation. Programs requiring many orbits get much greater scrutiny. The observations must address a significant astronomical mystery.

Proposals are divided into astronomical categories such as solar system objects, star formation, black holes in active galaxies, and the far universe. There are several types of proposals. The general observer (GO) proposals are the most common, with awarded telescope time typically ranging from 1 to 300 orbits. Snapshot observations—those for which targets require only 45 minutes or less of telescope time—are used to fill in gaps in the telescope schedule that cannot be filled by regular GO programs. Target of Opportunity proposals are for observation and analysis of unique astronomical events expected to occur during the year, such as the outburst of a variable star, or a newly discovered comet.

Up to 10% of telescope time is allocated by the Director of the Space Telescope Science Institute. Astronomers can apply to use this time throughout the year. It is typically awarded for study of unexpected phenomena such as supernovas. Historically, Institute directors have allocated large percentages of their time to special programs that are too big to be approved for any one astronomy team. This enabled the execution of the *Hubble* Deep Field (1996) and *Hubble* Ultra Deep Field (2004)—humankind's farthest views into the visible-light universe to date.

The *Hubble* and *Spitzer Space Telescopes* captured this image of the Orion Nebula in infrared, ultraviolet, and visible-light wavelengths. Located 1,500 light-years away from Earth, the Orion Nebula appears to the naked eye as a fuzzy "star" in the sword of the constellation Orion the Hunter. The largest and most technically challenging *Hubble* science program of 2008 was the 252-orbit program to observe protostars in this region.

Configuration, monitoring, and front-line troubleshooting of the network connections and components that comprise the *Hubble Space Telescope* Control Center System are accomplished around the clock by a small group of operators in the Data Operations Center at Goddard.

The proposal process is managed by the Institute. The GO proposals are peer reviewed annually by the Time Allocation Committee, which is broken up into selected disciplines, such as planetary science, stellar physics, and active galaxies. The committee looks for the best possible science that can be conducted by *Hubble* and seeks a balanced program of small, medium, and large numbers of orbits. The Institute Director approves the final selection and the winning Principal Investigators on the research teams are notified.

Because *Hubble* is not in continuous communication with the ground, a to-do list of observations is uploaded in packets several times a week for the telescope to automatically execute. Images and other data are stored in an onboard solid-state computer's memory and then downlinked to Earth at appropriate communications opportunities through NASA's Tracking and Data Relay Satellite System.

The ANGST Survey showed a swarm of young, blue stars in the diffuse dwarf irregular galaxy NGC 4163, which is a member of a group of dwarf galaxies located roughly 10 million light-years away.

After being received at Goddard Space Flight Center in Greenbelt, Maryland, the data stream into the Space Telescope Science Institute in Baltimore, where they are calibrated and archived. The Principal Investigator for the program receives an e-mail notice that his or her observation is complete and ready to be downloaded from the Institute's archive. *Hubble*'s astronomical observations are proprietary for only one year. After that time, other astronomers can view the data and conduct their own research.

One of the most important and interesting observations of 2008 was of the fading optical counterpart of a powerful gamma-ray burst that holds the record for being the most distant naked-eye object ever seen from Earth. For nearly a minute on March 19, 2008, this single "star" located 7.5 billion light-years away was as bright as 10 million galaxies combined. *Hubble*'s Wide Field and Planetary Camera 2 photographed the fading optical counterpart of the titanic gamma-ray burst, GRB 080319B, on April 7, 2008.

Astronomers had hoped to see the host galaxy where the burst presumably originated, but they were taken aback to find that the light from the gamma-ray burst was still drowning out the galaxy's light even three weeks after the explosion. Called a "long-duration gamma-ray burst," such events are theorized to be caused by the death of a very massive star, perhaps outweighing our Sun by as much as 50 times.

The largest and most technically challenging *Hubble* science program of 2008 was the 252-orbit program to observe protostars in the Orion region with the Near-Infrared Camera and Multi-Object Spectrometer. The program sought to understand the properties of the forming stars as well as the surrounding interstellar gas. The protostars were identified with the *Spitzer Space Telescope*, but *Hubble* data were needed to provide a more complete assessment of how the environment of the stars influences their formation and subsequent evolution.

Another large survey used 245 orbits to analyze 69 galaxies in the "Local Volume," a region that extends beyond our Local Group of galaxies. This region ranges from 6.5 million light-years to 13 million light-years from Earth. Under the leadership of Julianne Dalcanton of the University of Washington, this study, called the Advanced Camera for Surveys (ACS) Nearby Galaxy Survey Treasury (ANGST), collected detailed information on galaxy shapes while seeking to map their star formation histories.

The history of star formation in a galaxy can be determined by looking at the colors and brightnesses of the individual stars that make up the galaxy's light. From the ground, the light from the stars is smeared together, so a bright spot in a ground-based image could be due to a single bright star, or to dozens of fainter stars in a tight cluster. With *Hubble*'s resolution, the study's stars appear as distinct points.

Early results of the ANGST survey show the rich diversity of galaxies. The galaxies studied reside in a variety of environments, including close pairs, small and large groups, and isolated areas. An important finding is that the massive spiral galaxies—such as M81 in the study—formed in the early universe, confirming the results of previous galaxy surveys.

As the name implies, the ANGST survey was originally scheduled to use the ACS, the instrument responsible for many of *Hubble*'s most impressive images of deep space. Just 104 of the survey's orbits were completed when, in January 2007, the ACS stopped working because of an electrical short. By February 2007, one part of the instrument—the Solar Blind Channel—was returned to operation by reconfiguring the electrical system from the ground, but the limited capabilities of this channel were not sufficient to continue the ANGST observations.

Through flexibility and cleverness in scheduling, the survey was switched to the Wide Field Planetary Camera 2 instrument. Dalcanton and her group used 116 orbits of this instrument's time and then applied for Director's Discretionary time for an additional 25 orbits to reach their scientific goals.

As is standard, the images were temporarily stored in the solid-state memory onboard *Hubble*, and later downlinked via a NASA communications satellite to a ground terminal in White Sands, New Mexico. From there, the data were transferred to Goddard Space Flight Center, and finally to the Institute in Baltimore, where the images were stored in the *Hubble* data archive. At the same time, an automatic e-mail message sent to Dalcanton informed her that her group's images were available. These were provided as raw data sets so the researchers could conduct their own image processing, and also as institutionally processed images that were calibrated to remove any instrument artifacts in the data.

Even now, another batch of scheduled *Hubble* observations waits within the telescope's computers for execution at prescribed times. The coordinated efforts of many dedicated, detail-oriented people installed them in the queue. Many more people will work hard to ensure the observations are completed successfully, and then to collect, process, archive, analyze, and publish the results.

At peak levels, *Hubble* generates approximately 350 gigabytes of science data each month. Astronomers using these data have published more than 8,000 scientific papers, making *Hubble* the most productive space observatory ever built.

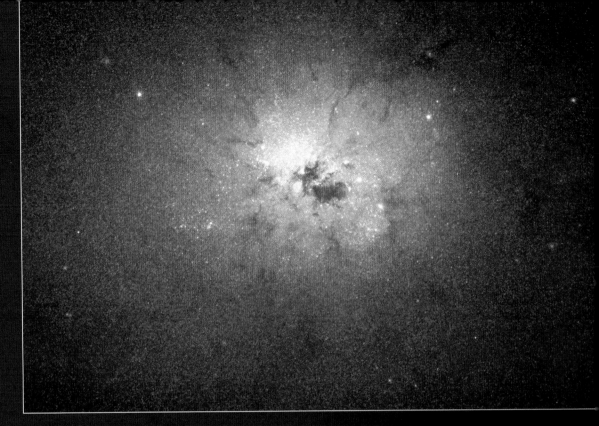

The ANGST survey revealed dark clumps of material scattered around the bright nucleus of NGC 3077—evidence of the galaxy's interactions with its larger neighbors. NGC 3077 is a member of the M81 group of galaxies and resides 12.5 million light-years from Earth.

Working under three, large, fixed-head star trackers, SM4 astronauts practice the installation of a rate sensor unit containing two gyroscopes. (Photo credit: M. Soluri/NASA)

SM4 Preparation

Servicing Mission 4 Preparation

With more than 19 years of ground-breaking science already accomplished, *Hubble* will be made fit for its final, culminating years of exploration when visited by Shuttle astronauts during Servicing Mission 4 (SM4), scheduled for 2009. Over the course of five spacewalks, the astronauts will install two new instruments, repair two inactive ones, and perform the component replacements that will keep the telescope functioning into at least 2014. If all goes according to plan, at the end of SM4 *Hubble* will be more powerful and capable than ever before.

Diverse groups of professionals across the country are busily preparing for the mission. Many activities involve upgrading ground systems, building new spacecraft hardware and tools, designing and optimizing operational procedures and timelines, and determining mission priorities and plans in the event of contingencies. They also include inspections, simulations, training, and orbiter preparations. The images on the following pages present a cross section of these activities. (Photo credit: M. Soluri/NASA except as otherwise noted)

SM4 Mission Patch

In order to prevent contamination from threatening the optical performance of *Hubble*'s scientific instruments, technicians use ultraviolet light to inspect the surfaces of flight hardware. Ultraviolet light aids in the detection of organic mineral deposits that remain on surfaces after more routine cleaning methods are complete.

With a rainbow serving as a backdrop in the sky, Space Shuttle *Atlantis* (foreground) sits on Launch Pad A and *Endeavour* on Launch Pad B at NASA's Kennedy Space Center in Florida. For the first time since July 2001, two shuttles are on the launch pads at the same time. *Endeavour* will stand by at pad B in the unlikely event that a rescue mission is necessary during Space Shuttle *Atlantis*' upcoming mission to service *Hubble*.

Space Shuttle *Endeavour*, the emergency back-up orbiter, is poised to leave the Vehicle Assembly Building (VAB) at Kennedy Space Center and roll out to Launch Pad 39B. The Shuttle stack, with solid rocket boosters and external fuel tank attached to the orbiter, rests on the mobile launcher platform and is moved by the crawler-transporter underneath.

The Space Shuttle *Endeavour* travels from the VAB to Launch Pad 39B. In the unlikely event that a rescue mission is needed during SM4, *Endeavour* will serve as the rescue vehicle and will be ready to launch.

The massive crawler-transporter carrying the mobile launch platform with the Space Shuttle, its external tank, and solid rocket boosters, travels at less than one mile an hour. The crawler travels on 8 tread belts, each containing 57 shoes. Each shoe is 7.5 ft long, 1.5 ft wide, and weighs approximately 2,100 pounds.

The payload canister is lifted into the payload changeout room above. The canister contains four carriers holding various *Hubble* equipment. The red umbilical lines keep the payload items environmentally controlled while in the canister. The changeout room is the enclosed portion of the rotating service structure that supports cargo delivery to the pad and subsequent vertical installation into the Shuttle's payload bay.

The crew of STS-125 arrive at the launch pad and visually inspect the enormous Space Shuttle, external tank, and boosters as part of a launch dress rehearsal known as the Terminal Countdown Demonstration Test. The test provides Shuttle crews the opportunity to participate in simulated countdown activities, including equipment familiarization and emergency training. From the left to right are Servicing Mission 4 astronauts Andrew J. Feustel, Michael J. Massimino, Michael T. Good, Scott D. Altman (commander), John M. Grunsfeld, K. Megan McArthur, and Gregory C. Johnson (pilot).

In the Orbiter Processing Facility at KSC, Servicing Mission 4 crew members participate in the Crew Equipment Interface Test activities. Here they are given experience in handling the tools, transport equipment, and *Hubble* flight hardware for the mission. Crew members are seen here being lowered inside the Space Shuttle's payload bay to inspect the Shuttle's robotic arm and orbiter boom sensor system.

Astronauts practice an emergency escape from the Shuttle as part of their training. Here they make their way from the orbiter to the slide-wire baskets that are on the fixed service structure designed to speed them away from the launch pad should a vehicle problem develop before launch.

Standing on suspended scaffolds that protect the orbiter's payload bay doors, technicians in the Orbiter Processing Facility at KSC inspect the cargo bay of *Atlantis* prior to its movement to the Vehicle Assembly Building where it will be integrated with its external tank and solid rocket boosters.

Astronauts Michael Massimino and Drew Feustel (sitting) practice repair techniques for the Space Telescope Imaging Spectrograph. This instrument failed on orbit well after its planned life expectancy period, but remains valuable enough to attempt fixing. The astronauts will attempt to replace the electronics card that contains a failed power supply—the first ever attempt at an instrument repair on orbit.

The flight crew participates in rehearsals of all of the mission's Extra Vehicular Activities (EVAs) as part of crew familiarization—shortened to "Crew Fam"—activities. On this day, astronauts Michael Good, Andrew Feustel, Michael Massimino and John Grunsfeld examine a flight tool container with Goddard engineers and *Hubble* servicing mission managers.

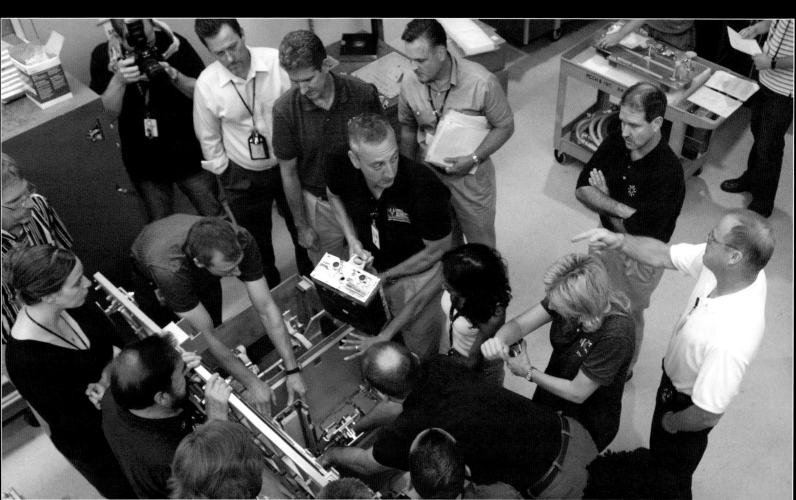

As part of their training, the astronauts periodically come to NASA's Goddard Space Flight Center in Greenbelt, Maryland to participate in specific Crew Fam activities. During these mission dress rehearsals, they have the unique opportunity of seeing all of the actual HST flight hardware in flight configuration in a clean room before they arrive on orbit together aboard the Space Shuttle *Atlantis*. This gives the astronauts the opportunity to work side-by-side with engineers from Goddard, as well as NASA's Johnson Space Center, to optimize hardware designs and EVA procedures.

Astronauts Andrew Feustel, Michael Good, John Grunsfeld, and Michael Massimino (left to right) pose for a picture after inspecting the Soft Capture Mechanism (SCM) mounted to the Flight Support System (background). The SCM is part of the Soft Capture and Rendezvous System (SCRS), which will enable the future rendezvous, capture, and safe disposal of *Hubble* by either a crewed or robotic mission. The SCRS enhances the capture interfaces on *Hubble*, therefore significantly reducing the rendezvous and capture design complexities associated with the disposal mission.

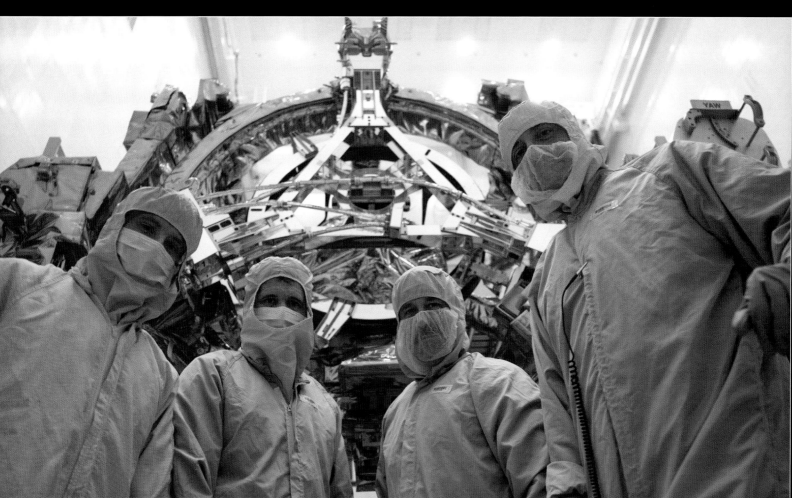

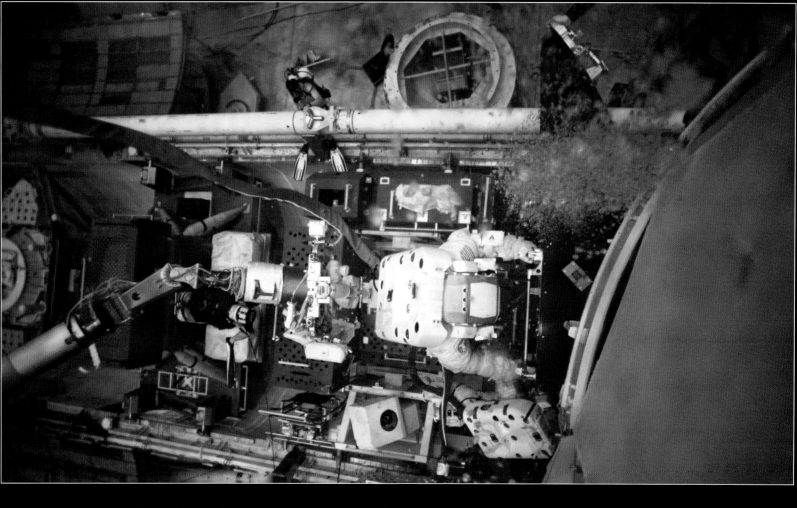

Astronauts spend many hours choreographing their procedures in order to be as efficient as possible while performing spacewalks to repair and upgrade the telescope. The Johnson Space Center Neutral Buoyancy Laboratory, a 6.2 million gallon pool, holds a full size mock-up of the Shuttle's cargo bay and *Hubble*, where these procedures are practiced. This is the closest the astronauts can get to practicing in a weightless environment. (Photo credit: NASA)

Practicing for gyro replacement, astronauts perform servicing steps while diver-engineers from the Goddard Space Flight Center closely assist to modify tools and processes as necessary. (Photo credit: NASA)

As seen from the top of the launch complex, Kennedy Space Center engineers work to configure the external fuel tank for launch. In the background, other launch support infrastructure is visible.

Astronauts Scott Altman and Greg Johnson participate in a cockpit simulation (right) along with their crewmates in Johnson Space Center's Mission Simulation and Training Facility. This standard exercise provides the astronauts the means to practice Shuttle landings under various mission scenarios.

Hubble News

Hubble observations have produced a regular stream of news about the universe. Shown here are a few recent highlights. Details on these topics and many others can be found on the World Wide Web at http://hubblesite.org.

Starburst Galaxy NGC 1569

Astronomers have long puzzled over why a small, nearby, and seemingly isolated galaxy is producing new stars faster than any other galaxy in our local neighborhood. *Hubble* has recently contributed important data that may solve the mystery. The data show that the galaxy, called NGC 1569, is one-and-a-half times farther away than astronomers originally thought.

The extra distance places NGC 1569 in the middle of a group of approximately 10 galaxies located some 11 million light-years away. Gravitational interactions among the group are thought to be compressing gas in NGC 1569 and igniting the star-making activity. The farther distance means that not only is the galaxy intrinsically brighter, but remarkably, that it is also producing stars twice as fast as originally thought. NGC 1569's star-forming rate is more than 100 times greater than the rate of our own Milky Way galaxy.

The *Hubble* study included observations of both NGC 1569's cluttered core and its sparsely populated outer fringes. Such studies are helping astronomers put together a more complete picture of the galaxies in the local universe.

Hubble showed that the galaxy NGC 1569 is one-and-a-half times farther away than astronomers originally thought.

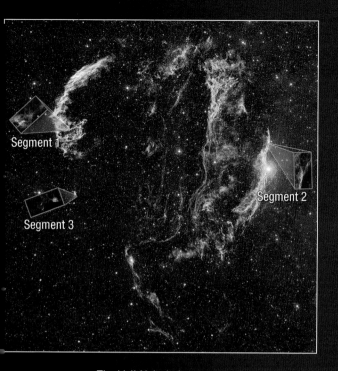

The Veil Nebula

Hubble photographed three magnificent sections of the Veil Nebula—the shattered remains of a supernova that exploded thousands of years ago. This series of images provides beautifully detailed views of the delicate, wispy structure resulting from this cosmic explosion. The Veil Nebula is one of the most spectacular supernova remnants in the heavens, stretching 75 light-years across. The entire shell spans about 3° on the sky, corresponding to the width of about six full moons. The explosion apparently obliterated the star, as no stellar remains have been found.

The Veil Nebula is a prototypical middle-aged supernova remnant, and is an ideal laboratory for studying the physics of supernova debris because of its unobscured location in our galaxy, its relative closeness, and its large size. Also known as the Cygnus Loop, the Veil Nebula is located in the constellation of Cygnus the Swan. It is about 1,500 light-years away from Earth.

The *Hubble* images are some of the best ever taken of optically emitting shock waves—places where a supernova blast wave is expanding into dense interstellar material. Variations in the density of the gas cause the shock speed to change, and the gas is ionized differently in various locations. These differences in ionization are reflected in the color variations. Scientists estimate that the supernova explosion occurred some 5,000 to 10,000 years ago. Observers at the time would have seen a star become bright enough to cast shadows at night.

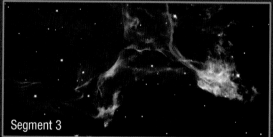

Views of the Veil Nebula. Above is a ground-based image from the Digitized Sky Survey; on the right are portions of the nebula as seen by *Hubble*. The Digitized Sky Survey was produced at the Space Telescope Science Institute and is based on photographic data from the Palomar Schmidt Telescope northern sky surveys (copyright Caltech) and the UK Schmidt Telescope southern sky surveys (copyright AATB/ROE).

A Star-Forming Region in the Large Magellanic Cloud

Revealed as a true celestial firestorm of birth and death, a small but dazzling portion of the nebula found near the star cluster NGC 2074, upper left, was recently brought to light by *Hubble*. A region of intense stellar creation thought to be triggered by a nearby supernova explosion, it lies about 170,000 light-years away in the Large Magellanic Cloud (LMC). A satellite of our Milky Way galaxy, the LMC is one of the most active star-forming regions in our Local Group of galaxies.

The 100 light-year wide landscape reveals dramatic ridges and valleys of dust, serpent-head "pillars of creation," and gaseous filaments glowing fiercely under torrential ultraviolet radiation. The region is on the edge of a dark cloud of molecular hydrogen that is an incubator for the birth of new stars.

High-energy radiation blazing out from clusters of hot young stars, already born in NGC 2074, is sculpting the wall of the nebula by slowly eroding it away through photoionization. The seahorse-shaped pillar at the lower right is approximately 20 light-years long, roughly five times the distance between our Sun and the nearest star.

The region is near a larger complex within the LMC known as the Tarantula Nebula, and is a valuable laboratory for observing star formation and evolution.

This region of intense stellar creation near the star cluster NGC 2074, upper left, is thought to be triggered by a nearby supernova explosion.

Ganymede is seen disappearing behind Jupiter as the moon moves about its seven-day orbit.

Jupiter and Ganymede

Composed of rock and ice, Ganymede is the largest moon in our solar system—larger in fact than even the planet Mercury. But the moon is dwarfed by its host planet Jupiter, the largest planet in our solar system—and a full 11 times larger than Earth.

In this *Hubble* image, Ganymede is seen disappearing behind the gas giant as it moves about the planet in its seven-day orbit. Because Ganymede's orbit is tilted nearly edge-on to Earth, the moon can routinely be seen passing in front of, and disappearing behind, its giant host. *Hubble*'s sharp view reveals features on Ganymede's surface, most notably the white impact crater, Tros, and its system of rays—the bright streaks of material blasted from the crater. The crater and its ray system are roughly the width of Arizona. The image also shows Jupiter's Great Red Spot, the large eye-shaped feature near the center of the picture. The size of two Earths, this storm has been raging for more than 300 years.

Astronomers are using this and other images obtained in timed sequences to study Jupiter's upper atmosphere. As Ganymede moves behind the planet, the sunlight it reflects passes through Jupiter's outer gaseous layers. Imprinted on that light is information about the properties of the high-altitude haze seen above the cloud tops.

A Star-Forming Region in the Milky Way Galaxy

Located approximately 20,000 light-years away, NGC 3603 is a prominent star-forming region in the Carina spiral arm of the Milky Way galaxy. One of the most massive young star clusters in the galaxy is imbedded in the nebula. This image, which spans roughly 17 light-years, shows the cluster—containing thousands of mostly blue stars—surrounded by a vast region of dust and gas. It reveals multiple stages in the life cycle of stars, and so is of particular interest to astronomers.

Powerful ultraviolet radiation and fast winds from the bluest and hottest stars are blowing an enormous bubble around the cluster. Moving into the surrounding nebula, this torrent of radiation sculpts the tall, dark stalks of dense gas, which are embedded in the walls of the nebula. These gaseous pillars are a few light-years tall and point to the central cluster. The stalks may be incubators for new stars.

On a smaller scale, a cluster of dark clouds called "Bok globules" resides at the top, right corner. First observed by astronomer Bart Bok in the 1940s, Bok globules are dark clouds of dense dust and gas in which star formation sometimes occurs. These clouds are about 10 to 50 times more massive than the Sun. Bok globules undergo gravitational collapse on their way to forming new stars.

The star-forming region NGC 3603 contains one of the most massive young star clusters in the Milky Way galaxy. On the right are enlarged views of (A) Bok globules and (B) a stalk of dense gas.

Globular Cluster M13

Globular clusters, found in a vast halo around our galaxy, appear to contain some of the oldest stars in the universe. Because they likely formed before the Milky Way's disk—where our solar system is located—they are older than nearly all the other stars in our galaxy. Studying these clusters helps astronomers understand the developmental history of the Milky Way.

Located 25,000 light-years distant, and approximately 150 light-years across, M13 is one of nearly 150 known globular clusters surrounding our galaxy. It is also one of the brightest, and contains over 100,000 stars. M13 is a favorite target for amateur astronomers, and can be glimpsed under dark skies with the naked eye in the spring constellation of Hercules.

The stars in M13, like in all globulars, collectively orbit the gravitational center of their spherically-shaped cluster. Near the core, the density of stars is about a hundred times greater than the stellar density in the neighborhood of our Sun. Evidence from *Hubble* suggests that stellar close encounters near the core can result in binary star systems—two stars orbiting each other—that rob one another of material and reignite as massive "new" blue stars.

Near the core of globular cluster M13, the density of stars is about a hundred times greater than the stellar density in the neighborhood of our Sun.

18 light-years

Ground-based image
(Photo Credit: T. Bash, J. Fox, and A. Block, NOAO/AURA/NSF)

Shells of Stars Around a Quasar

In concert with the W.M. Keck Observatory in Hawaii, *Hubble* has detected and characterized faint shells of stars around a quasar known as MC2 1635+119. Quasars are the extremely bright, active cores of galaxies that contain a central black hole—a region of space in which the gravity is so strong that nothing, not even light, can escape. MC2 1635+119 dominates the center of an elliptical galaxy located about 2 billion light-years away in the constellation Hercules. The shells are evidence that a collision with another galaxy occurred in the relatively recent past.

The quasar known as MC2 1635+119 and its host galaxy [center] against a backdrop of distant galaxies.

This clash funneled gas into the galaxy's center, feeding a supermassive black hole. The accretion onto the black hole is the source of the quasar's energy. This observation is the first time shells of stars have been observed in a quasar galaxy. Although shells have been studied for more than 20 years, this is the most spectacular case ever seen at such a distance. The observation supports the idea that most—if not all—quasars are born from interactions between galaxies.

Previous studies of this galaxy with ground-based telescopes showed a normal-looking elliptical galaxy containing an ancient population of 12-billion-year-old stars. With the superior resolution of *Hubble*'s Advanced Camera for Surveys the faint, thin shells were uncovered. Astronomers used the Keck Telescope to take the spectral signature of the light from the galaxy's stars. By modeling the spectrum, astronomers concluded that there are two distinct stellar populations in the galaxy—12-billion and 1.4-billion-year-old stars. The younger stars—combined with the existence of the shells—are evidence for the galactic merger.

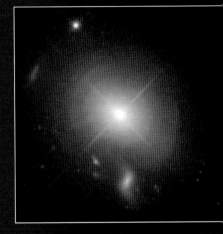

The shells are barely visible because of the bright light from the central quasar.

Hubble revealed at least five inner shells and additional debris. The shells resemble ripples in a pond when a stone is tossed into it. They formed when a passing galaxy was shredded by gravitational tidal forces as the two galaxies collided. Some of the first galaxy's stars were captured during that encounter and were then thrown into orbits that first carried them close to, and then later far from, the central nucleus of the elliptical galaxy. At their farthest point from the nucleus, these stars are moving most slowly along their orbits and tend to bunch up like cars in a traffic jam. These appear in the *Hubble* images as faint shells of starlight. As the trapped stars continue orbiting, they create successive shells, which give the illusion of expanding outward from the galaxy. The outermost shell is about 40,000 light-years from the galaxy's center.

This enhanced image reveals details of the faint shells.

A detail of the magnificent spiral galaxy M81 as seen by *Hubble*.

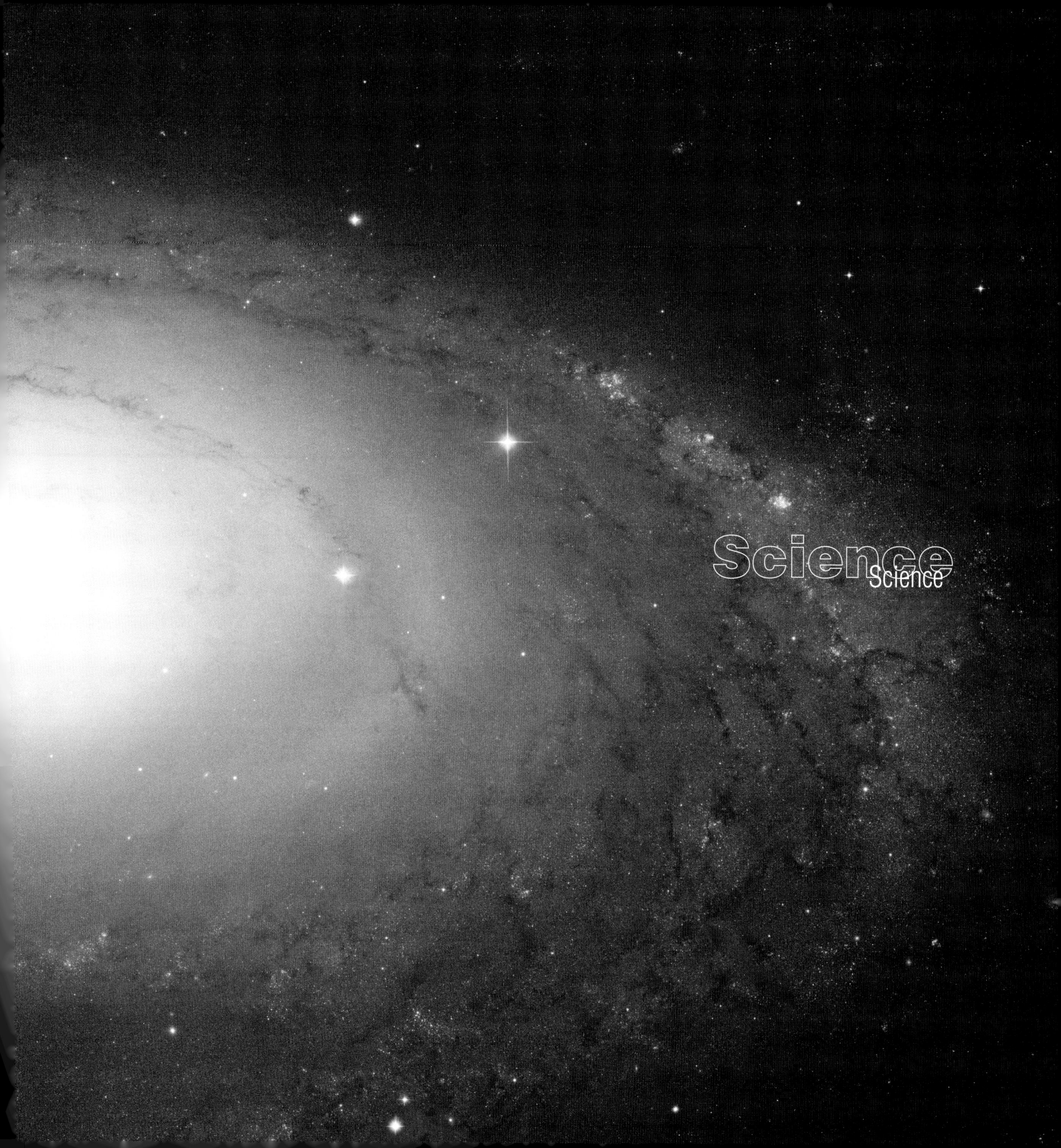
Science
Science

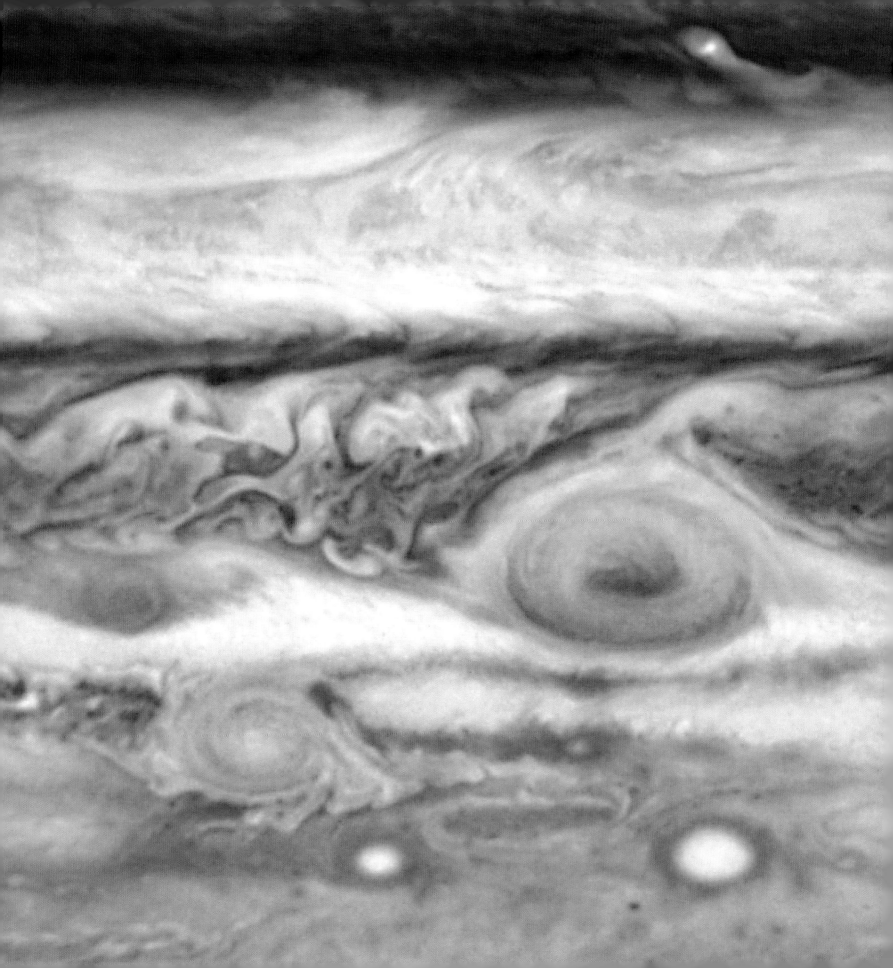

A New Red Spot on Jupiter

Since the mid-1600s, astronomers have marveled at changing spots and bands seen telescopically on the planet Jupiter. Larger, better telescopes revealed them to be dynamic features in Jupiter's dense atmosphere. The bands mark zones of prevailing winds, somewhat analogous to Earth's jet streams. The spots are cyclonic storms that rage along the turbulent boundaries of the jet streams. The largest of these, dubbed the Great Red Spot (GRS), is a cold, high pressure area 2–3 times wider than planet Earth. Its winds travel at approximately 400 miles per hour—far surpassing the 155 miles per hour wind speed of Earth's own Category 5 hurricanes.

In the 1930s, astronomers documented the collapse of a white jet stream just south of the GRS that resulted in the creation of three elongated features. During the next decade, astronomers watched these features morph into three distinct white cyclonic storms. In 1998, these storms merged together to form one giant white storm. Its wind speeds now rival those of the GRS and it has grown to nearly half its size. Then, abruptly in 2006—for reasons still not completely understood—that storm turned red. Astronomers named it "Red Spot Jr," or more formally, "Oval BA."

In 2008, yet another red spot developed. Named "Baby Red Spot," or "2008 Oval 2," this new storm was just a fraction of the size of the other two red spots. And unlike the others, its life span would be short—as it would be caught and absorbed within months by the anticyclonic spin of the GRS.

All three red spots were imaged in detail using *Hubble*, by two teams of astronomers led by Imke de Pater of the University of California, Berkeley and Amy Simon-Miller of NASA's Goddard Space Flight Center. *Hubble*'s first image of 2008 Oval 2 was taken in May 2008. The spot was located just to the west of the GRS and was at the same basic latitude. Oval BA was located between the two, although at a lower latitude.

On May 9 and 10, *Hubble* took this image, which revealed a new red spot on Jupiter (left-most spot). The Great Red Spot, larger than two Earths, can be seen to the right.

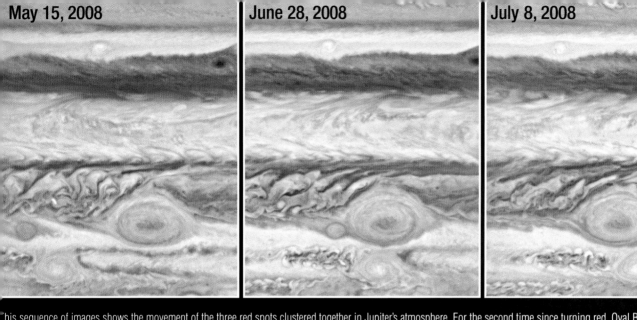

This sequence of images shows the movement of the three red spots clustered together in Jupiter's atmosphere. For the second time since turning red, Oval BA skirts past the Great Red Spot apparently unscathed. However, 2008 Oval 2 (identified by an arrow in the third panel) gets captured in the anticyclonic spin of the Great Red Spot.

The close alignment among the red spots was short-lived. *Hubble* images taken one and two months after 2008 Oval 2's discovery revealed that the small spot had traversed from the west side of the GRS to the east side. The small spot's appearance, however, was changed. The June and July images showed that 2008 Oval 2 was now trapped by the forces of the GRS storm. As it slipped into the maelstrom of the GRS, 2008 Oval 2 was disintegrating and losing its color.

No one is certain what gives these storms their red color. One theory is that the storms are so powerful that they dredge up gases or particulates from deep in Jupiter's atmosphere to higher altitudes, where ultraviolet radiation disassociates the elements within their molecules. Subsequent chemical reactions may lead to the red color. Astronomers speculate that the color loss in 2008 Oval 2 was due to the loss of its upward cyclonic circulation once it was trapped and consumed by the GRS.

Multi-wavelength images showed that the clouds in 2008 Oval 2 did indeed rise high into Jupiter's atmosphere before merging into the GRS. The cloud levels were documented in visible-light images captured by *Hubble* and by near-infrared data collected by the W.H. Keck Telescope in Hawaii. Astronomers determined that all three storms (the GRS, Oval BA, and 2008 Oval 2) appeared bright in near-infrared light, signaling that they were towering above the methane in Jupiter's atmosphere. Methane absorbs infrared light, so methane-rich planets like Jupiter appear dark in reflected sunlight.

As astronomers predicted, the GRS eventually consumed most of 2008 Oval 2, although some high clouds continued to move eastward after the encounter. A steady energy source of small storms, such as 2008 Oval 2, may be one reason why the huge GRS storm has sustained itself for so many years.

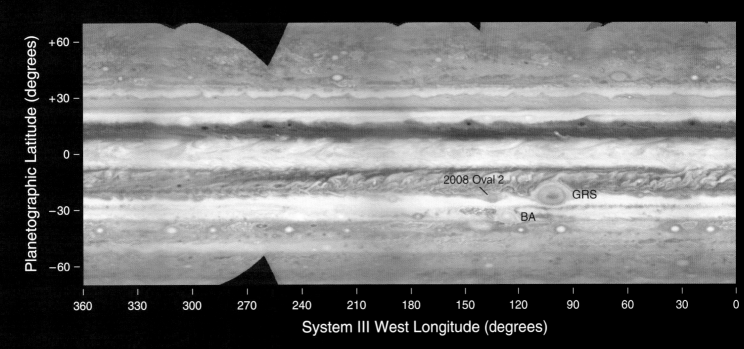

This false-color cylindrical map of Jupiter reveals the location of the newest red spot, 2008 Oval 2. Individual frames were obtained on May 9 and May 10, 2008 and mosaiced together to form this map with a resolution of a quarter degree in latitude and longitude. The red triangles at the top and bottom of the image are all areas of missing data.

The mighty planet Jupiter is 11 times the diameter of Earth, but rotates in less than half the time. Changing features within its turbulent atmosphere are noted and studied by amateur and professional astronomers alike.

Further Reading

Aguirre, E.L., "*Hubble* Zooms in on Jupiter's New Red Spot: The Great Red Spot's Baby Brother Falls Under the Space Telescope's Powerful Gaze," *Sky & Telescope*, **112**(2), 26–29, 2006.

Aguirre, E.L., "Jupiter's Mini Red Spot Survives Encounter," *Sky & Telescope*, **112**(6), 86–87, 2006.

Barrado-Izagirre, N., et al., "Jupiter's Polar Clouds and Waves from Cassini and HST Images: 1993–2006," *Icarus*, **194**(1), 173–185, 2008.

Cheng, A.F., et al., "Changing Characteristics of Jupiter's Little Red Spot," *Astronomical Journal*, **135**(6), 2446–2452, 2008.

McAnally, J.W., *Jupiter and How to Observe It*, London: Springer-Verlag, 2008.

Simon-Miller, A.A., et al., "Jupiter's White Oval Turns Red," *Icarus*, **185**(2), 558–562, 2006.

Amy Simon-Miller has observed Jupiter with the *Hubble Space Telescope* since 1993, including observations of the Great Red Spot in 2008. These data have been used to study the planet's ever-changing atmosphere in an effort to understand what powers the enigmatic storms and powerful winds. Born and raised in Union, New Jersey, Dr. Simon-Miller earned a B.S. in space sciences from the Florida Institute of Technology in 1993, and a Ph.D. in astronomy from New Mexico State University in 1998. She is Chief of the Planetary Systems Laboratory at the NASA Goddard Space Flight Center, where her focus is on understanding the atmospheres of Jupiter and Saturn and on designing future robotic space missions.

Imke de Pater is a professor in the departments of Astronomy, and Earth and Planetary Science at the University of California at Berkeley. Born in Hengelo, Netherlands, she received her Ph.D. cum laude in 1980 from Leiden University. She started her career observing and modeling Jupiter's synchrotron radiation, followed by detailed investigations of the planet's thermal radio emission. In 1994, she led a worldwide campaign at radio wavelengths to observe the impact of comet D/Shoemaker-Levy 9 with Jupiter. She is currently exploiting adaptive optics techniques in the infrared range to obtain high angular-resolution data of the volcanic activity on Io, the weather on Titan, planetary rings, and Jupiter's new red oval. She used the *Hubble Space Telescope* to study weather on Jupiter (in particular, the red Oval BA), and to complement adaptive optics observations of planetary rings with visible wavelength data. She teamed up with Jack Lissauer to write a book called *Planetary Sciences*, which is the graduate-level textbook in this field. (Photo credit: S. Anderson, Keck Observatory)

Dwarf Bodies in the Solar System

In 2006, the International Astronomical Union (IAU) rigorously debated the adoption of new definitions for the astronomical categories of "planet" and "dwarf planet." The Union's action was prompted by the discovery in the last decade of many significantly sized icy bodies in the Kuiper belt—the solar system's debris disk that extends outside the orbit of Pluto. Combined with the larger rocky bodies discovered long ago in the asteroid belt located between the planets Mars and Jupiter, there are now at least 125 planet-like objects known to be orbiting the Sun, with many thousands of smaller ones cataloged as well.

Given the new criteria, however, most of these bodies do not qualify as either planets or dwarfs. The IAU currently recognizes only eight planets and five dwarf planets. Ordered larger to smaller, the dwarf planets are Eris, Pluto, Haumea, Makemake, and Ceres.

NASA's *Dawn* spacecraft is en route to study the asteroid Vesta in 2011, and the dwarf planet Ceres in 2015. In preparation for the encounters, *Hubble* collected images of both these distant worlds. This important survey work, which includes searching for moons and debris around the objects, will make the *Dawn* mission more effective. Until *Dawn* arrives on location, *Hubble* has the best available resolving power to determine their shapes, and resolve their larger surface features as well. Findings from *Hubble*'s observations of these two bodies—along with the dwarf planet Eris—are detailed below.

This artist's concept depicts the dwarf planet Eris, the largest object found in orbit around the Sun since the discovery of Neptune in 1846. Like Pluto, Eris is a member of the Kuiper belt, a region of icy bodies beyond Neptune in orbit around the Sun. Eris is the largest known Kuiper belt object (KBO) to date, with Pluto ranking second.

Vesta

Named for the ancient Roman goddess of the hearth, Vesta was discovered in 1807 by Heinrich Olbers. About the size of Arizona, it is the second most massive and the third largest asteroid. Vesta revolves around the Sun in 3.6 terrestrial years and has an average diameter of about 330 miles (530 km). Residing in the main asteroid belt between Mars and Jupiter, it is large enough to be differentiated like Earth, with a crust, core, and mantle. Vesta also holds the distinction of being the brightest asteroid, and the only one ever visible to the naked eye.

Vesta is not a classified as a dwarf planet, however. This is because one of the newly adopted criteria for such status is that the body be spherical, or very nearly so (some flattening at the poles due to rotational effects, for instance, would not be a disqualifier). It must have sufficient gravity to pull the planetary mass into a shape of nearly constant radius throughout. While not true now, the discovery by *Hubble* of an impact crater on the asteroid's southern hemisphere nearly as wide as Vesta itself suggests that at one time Vesta did qualify—until a cataclysmic collision dramatically changed its shape.

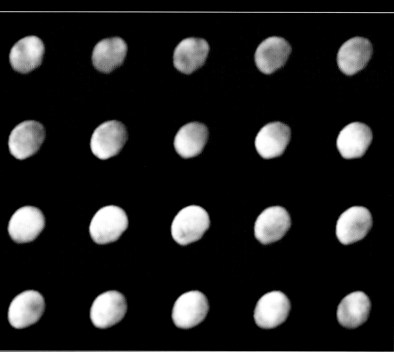

Astronomers took advantage of favorable Earth/Vesta geometry in 2007 and used *Hubble*'s Wide Field Planetary Camera 2 to collect images of the asteroid. *Hubble* in particular mapped the area dominated by the giant impact crater. The scar is 285 miles (456 km) across—evidence of a collision that blew approximately 1% of the asteroid's volume (over one-

To prepare for the *Dawn* spacecraft's visit to Vesta in 2011, astronomers used *Hubble*'s Wide Field Planetary Camera 2 to capture images of the asteroid. Despite Vesta's distance and diminutive size, *Hubble*'s optics resolved features as small as 37 miles (60 km) across. This is similar to reading the print on a golf ball from 25 miles away. These images help scientists to learn more about Vesta's surface structure and composition.

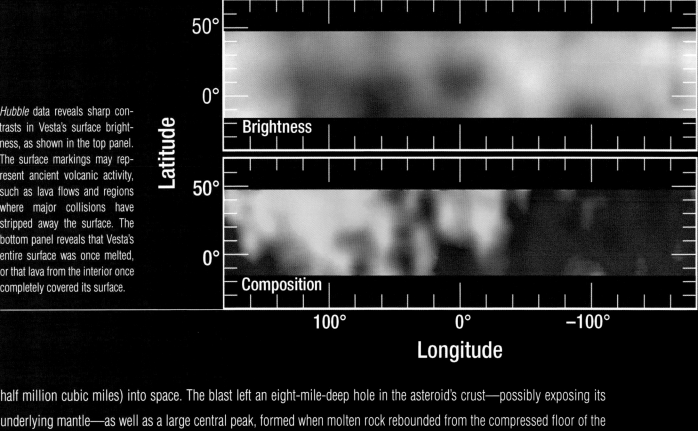

Hubble data reveals sharp contrasts in Vesta's surface brightness, as shown in the top panel. The surface markings may represent ancient volcanic activity, such as lava flows and regions where major collisions have stripped away the surface. The bottom panel reveals that Vesta's entire surface was once melted, or that lava from the interior once completely covered its surface.

half million cubic miles) into space. The blast left an eight-mile-deep hole in the asteroid's crust—possibly exposing its underlying mantle—as well as a large central peak, formed when molten rock rebounded from the compressed floor of the impact site. If Earth had a crater of proportional size, it would fill the Pacific Ocean basin.

The immense impact broke off large pieces of rock, producing more than 50 smaller asteroids that astronomers have nicknamed the "Vestoids." These are found throughout the inner main asteroid belt and are identifiable by their distinctive reflectance spectra, which match that of Vesta.

The impact crater is so large relative to Vesta's size that the collision might have been expected to cause more damage to the rest of the asteroid. Indeed, the crater lies near the asteroid's south pole, which is probably not coincidental. The excavation of so much material from one side of the asteroid would have shifted Vesta's rotation axis so that it settled over time with the crater near the pole.

In this *Hubble* picture of Vesta, a "nub" at the bottom of the asteroid marks the center of the giant crater and is suggestive of a catastrophic impact.

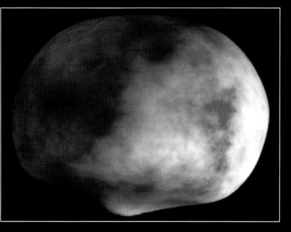

This three-dimensional computer model of the asteroid Vesta is synthesized from *Hubble* topographic data. The crater's eight-mile high central peak can be clearly seen near the south pole.

A color-encoded elevation map of Vesta clearly shows the giant 285-mile diameter impact basin and its central peak. Towering eight miles, this cone-shaped feature formed when molten rock rebounded from the crater's compressed center after the impact. The map was constructed from 78 pictures taken with the Wide Field Planetary Camera 2.

Unlike some other large asteroids that have jumbled surfaces due to a barrage of violent impacts, the rest of Vesta's surface appears to be largely intact. This view is derived from ground-based, spectroscopic measurements showing that Vesta has a surface of basaltic rock—frozen lava—which oozed out of the asteroid's presumably hot interior shortly after its formation 4.5 billion years ago. Heat during that period allowed heavier material to sink into Vesta's center, and lighter minerals to rise to the surface—a process known as planetary differentiation.

Hubble imagery resolved features as small as 37 miles (60 km) across, revealing widespread differences in brightness and color over the asteroid's surface. The brightness differences could be similar to those seen on the Moon. There, the smooth, dark regions are more iron-rich than the brighter highlands, which have minerals containing larger proportions of calcium and aluminum. Astronomers specifically combined images of Vesta in two colors to study the variations in iron-bearing minerals. The resulting view of the asteroid is that it has an intact, though variegated surface—something the *Dawn* spacecraft should see in great detail when it arrives there in August 2011.

This is the third time astronomers have turned *Hubble* toward Vesta. The first time, in 1994, *Hubble* obtained images covering a full rotation of the asteroid and were, therefore, able to map most of its surface area—about 200,000 square miles. Astronomers did not yet know about the giant crater, but they realized from Vesta's strange shape that there was something unusual about the asteroid. They had to wait for a better view from *Hubble* in 1996, when Vesta made it closest approach to Earth in a decade—only 110 million miles away. This is when the huge impact crater was clearly detected.

While Pluto was demoted to a dwarf planet, Ceres was promoted to the same category. Ceres is also an asteroid, and it resides with Vesta and tens of thousands of other asteroids in the main asteroid belt. The Texas-sized dwarf planet comprises about 30 to 40% of the asteroid belt's total mass.

Ceres revolves around the Sun in 4.6 years, and has a diameter estimated at about 590 miles. It is the largest asteroid, and was the first to be discovered. Named after the Roman goddess of agriculture, it was first observed in 1801 by Giuseppe Piazzi, who was looking for suspected planets in a large gap between the orbits of Mars and Jupiter. Gravitational perturbations from Jupiter billions of years ago prevented Ceres from accreting more material to become a full-fledged planet.

Hubble's very precise measurement of Ceres' limb revealed the dwarf planet's nearly round shape. This high degree of roundness suggests that Ceres' interior is layered like those of terrestrial planets such as Earth. The dwarf planet may have a rocky inner core, an icy mantle, and a thin, clay-like outer crust. It may even have water locked beneath its surface.

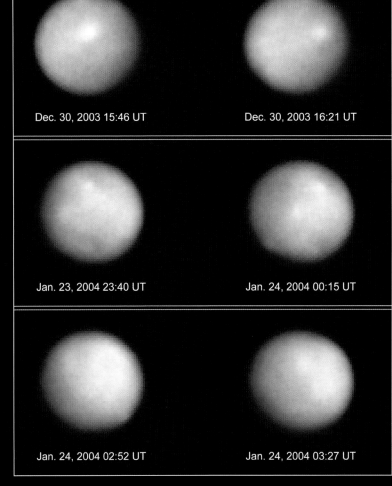

Hubble took these images of Ceres during its 9-hour rotation. Astronomers enhanced the sharpness in these *Hubble* Advanced Camera for Surveys images to bring out features on Ceres' surface, including brighter and darker regions that could be asteroid impact features.

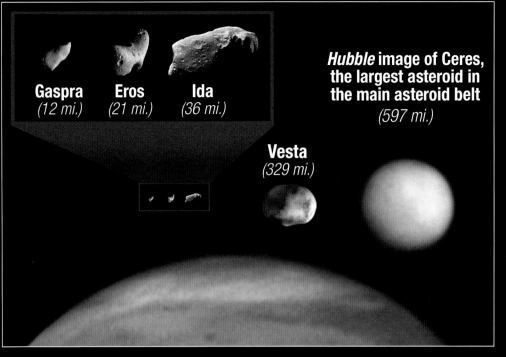

Ceres is compared with four other asteroids (Gaspra, Eros, Ida, and Vesta) and Mars.

Astronomers suspect that water ice may be buried under the crust because the density of Ceres is less than that of Earth's crust, and because the surface bears spectral evidence of water-bearing minerals. They estimate that if Ceres were composed of 25% water, it would contain more water than all the fresh water on Earth. Ceres' water, unlike Earth's, however, would be in the form of water ice located in the mantle, which wraps around the dwarf planet's solid core.

Ceres may have ice "volcanoes" or ice geysers. This hypothesis is based on theoretical modeling of Ceres thermal history, which suggests thermal temperature variations that may express themselves as geysers or volcanoes at the surface. From observations of the smooth surface at a scale of tens of kilometers, observers suggest that the surface may have been smoothed out by flowing water in the past.

Hubble also revealed bright and dark regions on Ceres' surface that could be topographic features, such as craters and/or areas containing different surface material. Large impacts may have caused some of these features and potentially added new material to the landscape. The color variations in the Hubble image show either a difference in the texture or the composition of the dwarf planet's surface.

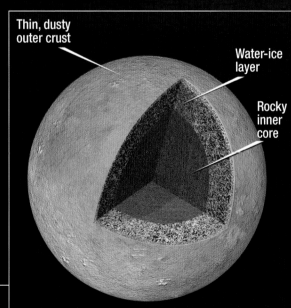

This cutaway view of Ceres shows the differentiated layers of the dwarf planet.

When Ceres and Vesta Were Planets

Pluto's dismissal from the planetary ranks is not unique. Ceres, Vesta, and the other asteroids found in the 1800s share similar stories. Guiseppe Piazzi at the Palermo Observatory found Ceres in 1801 in a gap between Mars and Jupiter where a planet was expected to reside, based on the spacing of the known planets in the solar system. So astronomers called it a planet.

A year after Ceres was discovered, astronomers found Pallas, another body between the same two planets, that was almost as bright as Ceres. Many astronomers realized that neither Ceres nor Pallas fit the conventional idea of a planet, because their disks were so small they could not be resolved through telescopes. Because of their star-like appearance, Sir William Herschel coined the term "asteroid" for such bodies. Most other astronomers, however, disagreed. These two new additions to the solar system were listed with the rest of the planets.

The list of small celestial bodies continued to grow. Astronomers identified Juno in 1804 and Vesta in 1807. These objects raised concern that the asteroids were debris from a planet that had somehow disintegrated. Nevertheless, Juno and Vesta joined Ceres and Pallas as planets. By the 1820s, astronomers counted 11 planets in the solar system. Introductory astronomy texts of that time listed the planets as Mercury, Venus, Earth, Mars, Vesta, Juno, Ceres, Pallas, Jupiter, Saturn, and Uranus.

By the end of 1851, there were 15 known bodies between Mars and Jupiter. Finally, astronomers realized that this large number of similar bodies, all in orbit between Mars and Jupiter, represented a new class of solar system object. They called them asteroids, the name Herschel had coined 50 years earlier. Astronomers today list about 100,000 known asteroids at least as large as 6 miles (10 km) across located between Mars and Jupiter, a region now called the asteroid belt.

When NASA's *Dawn* spacecraft reaches Ceres in 2015, its close-proximity imaging will help unravel the true cause of the dwarf planet's varied surface brightness. Meanwhile, *Hubble* imagery is the best available, and will help the *Dawn* mission team formulate the best plan for conducting its investigation.

This image of the dwarf planet Eris (center) and its satellite Dysnomia (near the 9 o'clock position) was taken with *Hubble*'s Advanced Camera for Surveys on August 30, 2006. *Hubble* images were combined with images from the Keck telescopes to measure the satellite's orbit and calculate a mass for Eris.

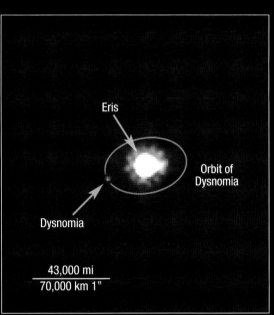

This labeled *Hubble* image of Eris and Dysnomia shows the satellite's orbit around the dwarf planet.

Eris

The dwarf planet Eris is named for the goddess of warfare and strife. According to Greek mythology, she stirs up jealousy and envy to cause anger and fighting among men. In the astronomical world, Eris produced trouble internationally when the question of its proper designation led to a contentious debate at the 2006 IAU meeting in Prague. If *Hubble* had not determined that the dwarf planet Eris was actually about 5% larger than Pluto, Pluto might have retained its status as a full-fledged planet—at least for a little while longer. The IAU had to decide whether to instate Eris as the tenth planet, or categorize both Pluto and Eris as dwarf planets. In the end, the decision was made to demote Pluto, and the solar system was left with only eight planets.

Eris is the largest object found in orbit around the Sun since the discovery of Neptune in 1846. Like Pluto, this dwarf planet is a member of the Kuiper belt, a region of icy bodies beyond Neptune in orbit around the Sun. It is the largest known Kuiper belt object (KBO) to date, with Pluto ranking second. Eris is three times more distant than Pluto and takes more than twice as long to orbit the Sun. Its orbital eccentricity—how "round" or elliptical the orbit is in shape—exceeds that of Pluto. Over the course of its 250-year orbital trek, Pluto varies in its distance from the Sun from 30 to 50 astronomical units (AU—the distance from Earth to the Sun—approximately 93 million miles). Eris moves from 38 to 97 AU over a 560-year orbit.

With large Earth-bound telescopes, Eris is too small to be seen as anything but a dot of light. Using *Hubble*, however, astronomers were able to directly measure its angular diameter. Eris is 1491 miles across (2400 km), plus or minus 62 miles (100 km).

Planet vs. Dwarf Planet

Pluto's planetary pedigree was put to the test when astronomers began finding other icy, rocky bodies throughout the Kuiper belt. *Hubble* observations determined that one of the objects, Eris, is even larger than Pluto. If Pluto were to remain a planet, Eris would have to join the planetary ranks as well.

In 2006, the International Astronomical Union (IAU), an astronomers' professional society, attempted to settle the debate by adopting a new definition for the word "planet." The new IAU definition states that in the solar system, a full-fledged planet is a celestial body that: (1) is in orbit around the Sun; (2) has sufficient mass so that it takes on a spherical shape; and (3) has "cleared the neighborhood" around its orbit by the action of its gravitational field. A body fulfilling only the first two of these criteria is classified as a "dwarf planet."

Joint observations using *Hubble* and the W.M. Keck Observatory determined its mass. This was calculated by observing the orbital period of Eris' moon Dysnomia as it moved around the dwarf planet. Using multiple images from both telescopes, astronomers determined that Eris has 1.27 times the mass of Pluto.

Dysnomia is in a nearly circular 16-day orbit, which favors the idea that this moon was born out of a collision between Eris and another Kuiper belt object. A gravitationally captured object would likely be in a more elliptical orbit. The satellites of Pluto, as well as the Earth–Moon system itself, are believed to have formed from collisions.

Using its mass and diameter, astronomers have calculated a density for Eris of 2.3 grams per cubic centimeter. This is very similar to the values determined for Pluto, Neptune's moon Triton, and the large KBO designated 2003 EL61. Densities higher than 1 gram per cubic centimeter imply that these bodies are not pure ice, but must have a significantly rocky composition.

Very deep *Hubble* observations indicate that Eris has just one satellite. Astronomers will soon turn *Hubble* toward Vesta and Ceres to look for moons around these bodies as well. It is very possible that Vesta might have at least one moon, which could have coalesced out of debris from the enormous impact to its southern region.

No doubt the *Dawn* mission to Vesta and Ceres will make many new discoveries about these diminutive members of the solar system. Meanwhile, with its unrivaled resolution, *Hubble* is uniquely suited to characterize these and other dwarf bodies in the solar system and thus, will help astronomers understand and explore these miniature, but fascinating, worlds.

Further Reading

Adler, J., "Of Cosmic Proportions: Astronomers Decide Pluto Isn't a Real Planet Anymore. Why They Did It—And How Our View of the Universe Is Changing," *Newsweek*, September 4, 2006.

Barucci, M.A., ed., *The Solar System Beyond Neptune*, Tucson: University of Arizona Press, 2008.

Brown, Michael E., et al., "Direct Measurement of the Size of 2003 UB313 from the Hubble Space Telescope," *The Astronomical Journal*, **643**(2), L61-L63, 2006.

Brown, M.E., and E.L. Schaller, "The Mass of Dwarf Planet Eris," *Science*, **316**(5831), 1585, 2007.

The Johns Hopkins University, Applied Physics Laboratory, "The Great Planet Debate: Science as Process, A Scientific Conference and Educator Workshop," URL http://gpd.jhuapl.edu/, August 14–16, 2008.

Thomas, P.C., et al., "Differentiation of the Asteroid Ceres as Revealed by Its Shape," *Nature*, **437**, 224–226, 2005.

van der Hucht, K.A., ed., *Transactions IAU, Volume XXVIB, Proceedings, IAU XXVI General Assembly*, August 2006, Cambridge, England: Cambridge University Press, 505 pp., 2008.

Weintraub, D.A., *Is Pluto a Planet? A Historical Journey through the Solar System*, Princeton, N.J., Princeton University Press, 2007.

Lucy McFadden's interest in Vesta dates back to her days as a graduate student. With her thesis advisor, she measured the asteroid's reflectance spectrum, which carries a mineralogical fingerprint of its surface rocks. Now as Co-Investigator and Director of Education and Public Outreach for NASA's *Dawn* mission, she is preparing for *Dawn*'s visit to Vesta and Ceres by developing compositional maps and conducting deep searches for satellites around these asteroids. Born in New York City, she earned a B.S. from Hampshire College in Amherst, Massachusetts with a concentration in astronomy and geology. She received her M.S. in Earth and planetary sciences from MIT, and her Ph.D. in geology and geophysics from the University of Hawaii. She serves as a Research Professor in the Department of Astronomy at the University of Maryland at College Park, and her career as a planetary scientist involves studying the formation of solar system bodies. She has developed and managed programs both at the University of Maryland and in her local community to inspire in students a love of learning and to develop the skills needed to pursue careers in science, technology, engineering, and math.

Joel William Parker headed the effort to study Ceres with *Hubble* in 2005. These observations led astronomers to believe the dwarf planet may have a differentiated interior and may contain large amounts of water ice beneath its surface. Born and raised in San Leandro, California, Dr. Parker earned a B.S. in physics and astronomy from the University of California at Berkeley in1986, and a Ph.D. in astrophysics from the University of Colorado at Boulder in 1992. He is the Assistant Executive Director at Southwest Research Institute in Boulder, where his focus is on Kuiper belt objects, comets, and asteroids. He produces and hosts the science show "How on Earth" on KGNU radio and has been involved in theatre and film for nearly 30 years.

Mike Brown is best known for his discovery of Eris, the largest object found in the solar system in 150 years. Born in Huntsville, Alabama, he grew up listening to the test firing of Saturn V rockets. He received his A.B. from Princeton in 1987, and his M.A. and Ph.D. from the University of California at Berkeley, in 1990 and 1994, respectively. He is a professor of planetary astronomy at the California Institute of Technology, where he has been on the faculty since 1996. He specializes in the discovery and study of bodies at the edge of the solar system. In 2006, he was named one of *Time* magazine's 100 Most Influential People.

First Visible-Light Image of an Extrasolar Planet

When the *Hubble Space Telescope* was launched in 1990, one of its most ambitious goals was to take a snapshot of a planet orbiting another star. Astronomers have now achieved that dream, capturing the first visible-light image of an extrasolar planet. Called Fomalhaut b, the planet orbits the bright southern star Fomalhaut, located 25 light-years away in the constellation Piscis Australis, the Southern Fish. It is estimated to be no more than three times the mass of Jupiter.

Exoplanets are usually detected indirectly, either spectroscopically by the subtle wobble of a star as it is tugged by the gravity of an unseen giant planet, or by the slight dimming of the star's light as its planet passes in front of it as seen from Earth. In this case, for the first time ever, the planet was directly seen by the astronomers as a distinct object in visible-light images. Observations taken 21 months apart by *Hubble*'s Advanced Camera for Surveys' coronagraph show that Fomalhaut b is moving along a path around the star, and therefore, is gravitationally bound to it. The planet is 10.7 billion miles from its star, or about three times the distance of the planet Neptune from the Sun.

Following the Dust
The star Fomalhaut has been a candidate for planet hunting ever since an excess of dust was discovered around it in the early 1980s by NASA's *Infrared Astronomy Satellite* (IRAS). In 2004, astronomer Paul Kalas, of the University of California at Berkeley, and his team members used the coronagraph on *Hubble*'s Advanced Camera for Surveys to view Fomalhaut. (A coronagraph blocks the light of a bright star, removing most of the glare so astronomers might see the faint light from planets.) The team found a vast dust belt circling the star—an analog to the Kuiper belt, which is a system of icy bodies in our own solar system that range from dust grains to objects the size of dwarf planets, such as Pluto.

This artist's concept shows the newly discovered planet, Fomalhaut b, orbiting its star, Fomalhaut. The exoplanet may have a structure similar to the Saturn-like ring seen in the foreground and encircling the planet. The star Fomalhaut also is surrounded by a ring of material, which appears as the bright, diagonal line in the background. (Figure credit: L. Calcada/ESO)

While much larger than our solar system's Kuiper belt, the *Hubble* image clearly showed that this structure is a ring of debris. Approximately 21.5 billion miles across, the belt reflects various frequencies of the star's light into space, but notably optical ones where *Hubble* detected it—a historic observation of its own. But it also had several unusual features believed by the team to be due to the gravitational influence of a planet not yet seen.

The first was the sharp inner edge of the ring. This is consistent with the presence of a planet that gravitationally "shepherds" ring particles out of one zone and into another—much as tiny moons in the Saturn ring system are known to do. The second was that the center of the ring appeared offset from the center of the star. This too is a telltale sign of an unseen body influencing ring material out of a round orbit into one that is more eccentrically shaped. The team used these observations to predict the location of a planet within Fomalhaut's ring system.

At the time, they noted a few bright sources in the image as potential planet candidates. Returning to Fomalhaut in 2006 with *Hubble*, they obtained even deeper and more accurate images of the belt. These images showed that one of the objects is moving through space with Fomalhaut, but changed position relative to the dust ring since the 2004 exposure. This object, Fomalhaut b, is orbiting the star Fomalhaut just inside the dust belt, thereby satisfying the team's hypothesis that the belt is heavily influenced by the gravity from this planet. The planet lies 1.8 billion miles inside the belt's inner edge. Fomalhaut b not only cuts out a sharp inner edge to the belt, but it also produces an entire shift of the belt's center away from the star. Fomalhaut b completes an orbit in roughly 872 years.

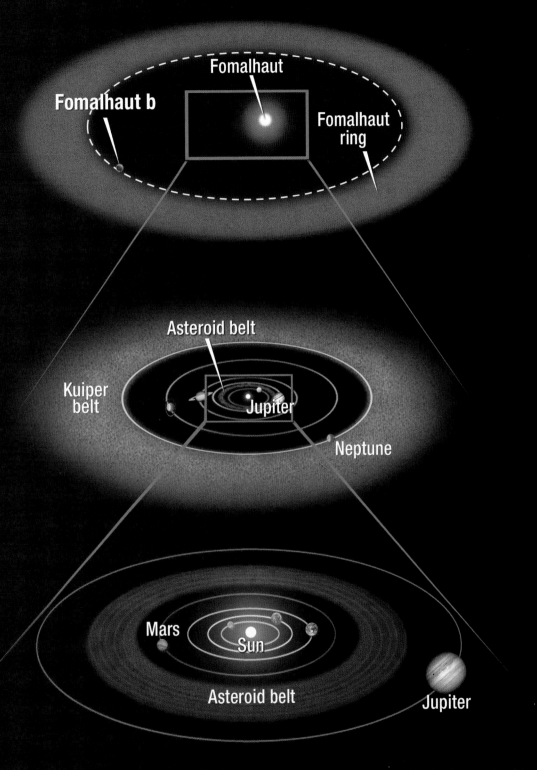

Our entire solar system, including the Kuiper belt, would fit comfortably inside Fomalhaut b's orbit. Fomalhaut b sculpts the inner edge of the circumstellar disk in much the same way that Neptune sculpts the Kuiper belt.

But Is It Really a Planet?

The astronomers used two techniques to come to the same conclusion: Fomalhaut b is indeed a planet, and not some other type of object in orbit around Fomalhaut. First, they measured its brightness—one billionth the brightness of its central star. If Fomalhaut b were a brown dwarf or a star, it would have been much brighter and more easily detectable. From the faintness of Fomalhaut b, the team is confident that it is smaller than three Jupiter masses, and thus planet-sized.

The other technique involved modeling the gravitational effects of different-sized masses on a dust belt encircling a star. Prior computer simulations showed that circumstellar disks in general—not just Fomalhaut's—will be gravitationally modified if one or more unseen planets are present and have certain masses and distances from the belt. By measuring how close it is to the dust belt and running the model, the team estimated the mass of the object. If Fomalhaut b were very massive, its gravity would have severely disrupted the dust belt. In fact, the model calculated that for the belt to be preserved in its currently observed state, the object would have to be less than three Jupiter masses—again clearly qualifying it as a planet. It took the science team four months of analysis and theoretical modeling to reach this conclusion.

The team hopes to obtain future observations to track the planet as it moves along its orbit. By obtaining more positions spaced over time, they will be able to determine the actual orbit of Fomalhaut b to greater accuracy. As the orbit becomes better known, they will gain more insight into Fomalhaut b's mass and its possible evolution.

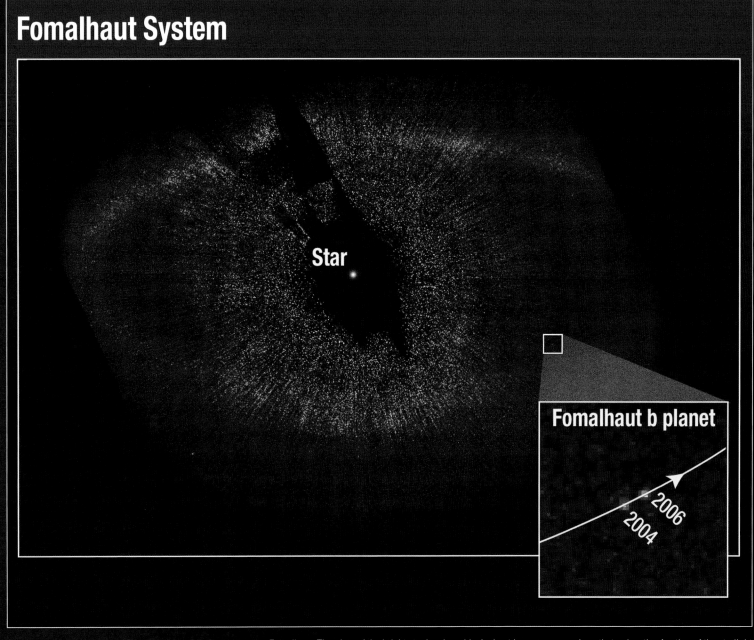

The newly discovered planet, Fomalhaut b, orbits its parent star, Fomalhaut. The glare of the bright star has been blocked out by a coronagraph so that only a tiny fraction of the starlight bleeds through. The small, white box at the right pinpoints the planet's location. The inset at bottom right is a composite image showing the planet's position during *Hubble* observations taken in 2004 and 2006.

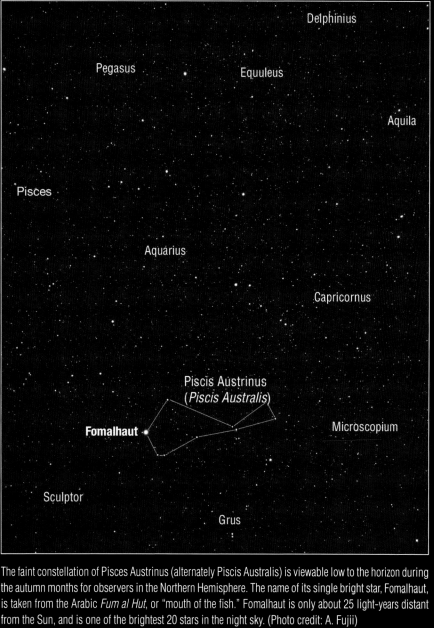

The faint constellation of Pisces Austrinus (alternately Piscis Australis) is viewable low to the horizon during the autumn months for observers in the Northern Hemisphere. The name of its single bright star, Fomalhaut, is taken from the Arabic *Fum al Hut*, or "mouth of the fish." Fomalhaut is only about 25 light-years distant from the Sun, and is one of the brightest 20 stars in the night sky. (Photo credit: A. Fujii)

Mysteries of Fomalhaut b

Fomalhaut, a type A star, is bigger, brighter, hotter, and younger than the Sun. The star is about 2.3 times the mass of the Sun and is about 200 million years old. The Sun is about 4.5 billion years old. One of the 20 brightest stars in the sky, Fomalhaut is easily visible with the naked eye.

Because the Fomalhaut system is only 200 million years old, the planet should be a bright infrared object still cooling through gravitational contraction. By modeling the planet's atmosphere, the researchers expected to be able to see Fomalhaut b in infrared light, to detect the heat from the planet. But even with using large ground-based telescopes at infrared wavelengths, they have not yet detected the planet. This sets an upper limit on its mass, because the bigger the planet, the hotter and brighter it would be (at its given age).

On the other hand, Fomalhaut b was fairly easily seen at optical wavelengths, where thermal emission should be weaker. Furthermore, the planet was seen by *Hubble* to have a "bluer" color in the optical than would have been expected from the models of

One possibility for the brightness properties of Fomalhaut b is that it has a huge, Saturn-like ring of ice and dust reflecting light from the central star (Fomalhaut itself). This disk might eventually coalesce to form moons. The ring—if it exists—has an estimated size comparable to the region around Jupiter that is filled with the orbits of Jupiter's four largest moons. The disk could be similar to what Jupiter looked like when it was 100 million years old and was surrounded by a dust disk out of which these moons formed. Therefore, the researchers believe that *Hubble*'s optical detection of the planet could have been greatly aided by reflection of Fomalhaut's light from a ring system.

If Fomalhaut b does have a disk of gas and dust around it that is producing moons, then that in itself is a mystery. The astronomers believe that any planet around the star Fomalhaut should already have formed its moons at this age and largely lost its disk or rings. They are not sure why such a disk of material around a planet would persist for 200 million years— opening new avenues for additional theoretical work.

Another mystery is that the planet unexpectedly dimmed in brightness by about 40% between the 2004 and 2006 observations. This might mean that it has a hot outer atmosphere heated by bubbling convection cells on the young planet. Or, it might come from hot gas at the inner boundary of a ring around the planet. It is just too early to tell exactly what is happening in Fomalhaut b, but the current mysteries offer plenty of reasons for additional observations of this fascinating object.

Comparison to Our Own Solar System

Some similarities exist between our own solar system and the Fomalhaut system, but there are also many differences. Fomalhaut is a brighter star than our Sun—16 times brighter—and its system is on a much larger scale than our own solar system. The planet Fomalhaut b is orbiting around the inside of a dust belt and dynamically sculpting and shaping it. In our own solar system, Neptune operates similarly, sculpting and shaping the inside edge of the Kuiper belt.

The planet may have formed at its location in a primordial circumstellar disk by gravitationally sweeping up remaining gas. Or, it may have migrated outward through a game of gravitational billiards, where it exchanged momentum with smaller planetary bodies. It is commonly believed that the planets Uranus and Neptune migrated out to their present orbits after forming closer to the Sun and then moving through gravitation interaction with smaller bodies.

Fomalhaut is significantly hotter than our Sun, which means a planetary system could scale up in size with a proportionally larger Kuiper belt feature and scaled-up planet orbits. For example, the "frost line" in our solar system—the distance where ices and other volatile elements will not evaporate—is roughly at 500 million miles from the Sun; but for hotter Fomalhaut, the frost line is at roughly 1.9 billion miles from the star.

Fomalhaut is burning hydrogen at such a furious rate through nuclear fusion that it will burn out in only 1 billion years, which is one-tenth the lifespan of our Sun. This means the opportunity for advanced life to evolve on any habitable worlds the star might possess is much lower than here on Earth.

Future Observations

With the refurbishment of the *Hubble Space Telescope*, Kalas and his team plan to keep observing Fomalhaut b and refining its orbit. They also hope to obtain a spectrum, which is the only way in the near future to definitively determine if there is a ring around the planet. This is very difficult to accomplish because of the planet's faintness, but should be within the refurbished *Hubble*'s reach.

Using NASA's *James Webb Space Telescope* (*Webb*), scheduled to launch in 2013, astronomers will be able to make coronagraphic observations of Fomalhaut b in the near- and mid-infrared. *Webb* will also be capable of looking for evidence of water vapor clouds in the planet's atmosphere. This will offer clues to the evolution of a comparatively newborn 100-million-year-old planet.

Webb will also be able to hunt for other planets in the system and probe the region interior to the dust ring for structures such as an inner asteroid belt. Our own solar system has a rich diversity of terrestrial planets, gas giant planets, and additional dust structures. Kalas and his team will look for these same structures in the Fomalhaut system. They view Fomalhaut b as a Rosetta Stone that will enable the decoding of future images of disks like Fomalhaut's and allow astronomers to find other planets using similar techniques.

Further Reading

Horne, K., "The Quest for Extra-Solar Planets," *AIP Conference Proceedings*, **848**(1), 787–799, 2006.

Kalas, P., et al., "Optical Images of an Exosolar Planet 25 Light-Years from Earth," *Science*, **322**(5906), 1345–1348, 2008.

Kalas, P., J.R. Graham, and M. Clampin. "A Planetary System as the Origin of Structure in Fomalhaut's Dust Belt," *Nature*, **435**(7045), 1067, 2005.

Naeye, R., "The Planet-Disk Connection," *Sky and Telescope*, **110**(4), 18, 2005.

Santos, N.C., "Extra-Solar Planets: Detection Methods and Results," *New Astronomy Reviews*, **52**(2-5), 154–166, 2008.

Dr. Paul Kalas is known for his discoveries of debris disks around stars. He was born in New York City and raised in Detroit, Michigan. He earned his B.S. in astronomy and physics in 1989 at the University of Michigan in Ann Arbor, and his Ph.D. in astronomy in 1996 from the University of Hawaii at Manoa. In 2006 he became an adjunct professor of astronomy at the University of California at Berkeley. Using the *Hubble Space Telescope*, Kalas led a team of scientists to obtain the first visible-light images of Fomalhaut b, an extrasolar planet around the star Fomalhaut. Dr. Kalas also discovered several circumstellar disks using the *Hubble* telescope's coronagraph and the University of Hawaii's 2.2-meter telescope at Mauna Kea. In 1995, he discovered various forms of asymmetric structure in optical images of the Beta Pictoris disk. He was the lead scientist for the first optical images of debris disks surrounding the nearby red dwarf star AU Microscopii and the bright star Fomalhaut.

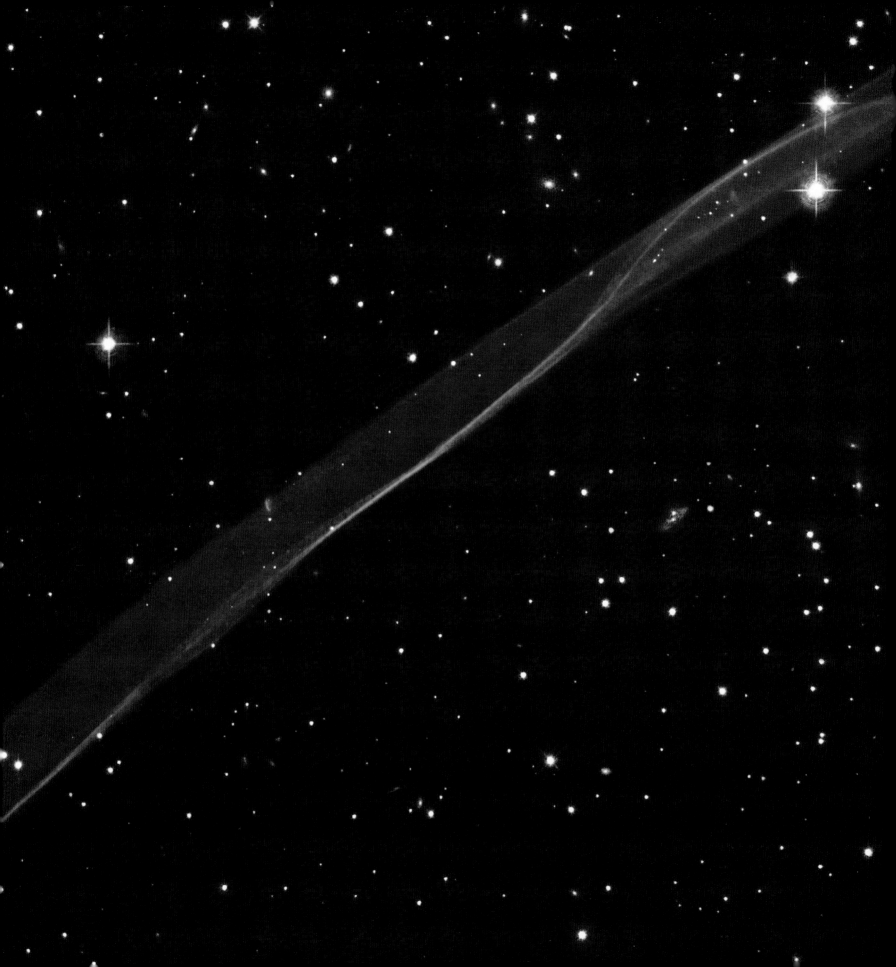

Supernova Remnant SN 1006

More than 8,000 years ago, the cataclysmic explosion of a nearby star sent a shock wave racing outward through space at nearly 20 million miles an hour. Found in what is now the southern constellation of Lupus, the Wolf, the explosion—a supernova—completely annihilated the star. The shock wave heated all the interstellar gas and dust in its path. Today, a delicate, red-colored "ribbon" of gas appears to be floating alone in space in the direction of the former star. This "ribbon," in reality, is just a small portion of a large bubble of material called a supernova (SN) remnant that surrounds the site of the explosion.

Known as SN 1006, the spherically-shaped stellar remnant is invisible at optical wavelengths except for this ribbon-like structure. Radio telescopes discovered its presence in the 1960s, detecting a nearly circular ring of material at the location of the supernova. In 1976, astronomers used the 4-m Blanco telescope at the Cerro Tololo Inter-American Observatory in Chile to observe a faint optical glow from a small region on the northwest edge of the radio ring. A small slice of that region is shown in detail in the *Hubble* image. The thin strand corresponds to a part of space where the shock wave is sweeping up tenuous hydrogen gas, heating it, and causing it to radiate in visible light.

The twisting ribbon is indeed an optical illusion. It appears ribbon-like because it is observed almost exactly along the edge of the expanding bubble. The strand is actually more like a crinkled sheet of paper or cloth viewed from the side. Slight ripples in the sheet produce the sharp edges, and more diffuse light fills in between them.

Over the past few decades, astronomers have made observations of the SN 1006 remnant in many different wavelengths of light. The presence of optical emission along only one section of this huge, expanding bubble tells astronomers that neutral hydrogen gas is present in significant quantities only along this portion of the shell.

> This very thin section of a supernova remnant appears as a ribbon because *Hubble* is looking almost exactly along the edge of the expanding bubble, an object approximately 60 light-years across. The remnant was the product of a stellar explosion that occurred more than 8,000 years ago, but was only seen by observers in the year 1006 A.D., as its light took 7,000 years to travel the distance to Earth.

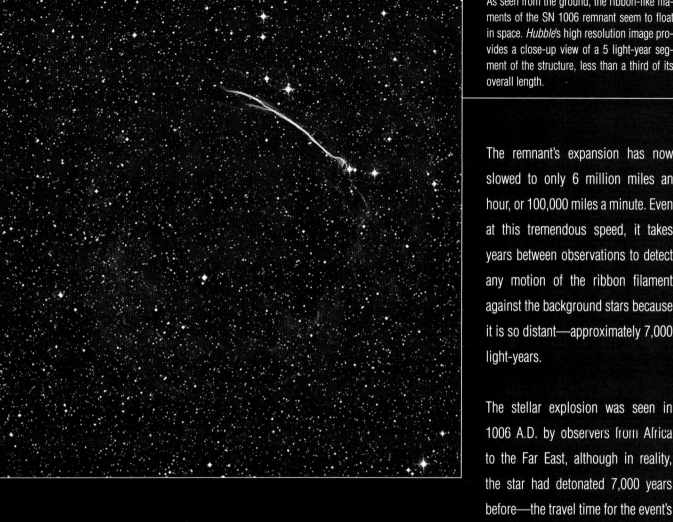

As seen from the ground, the ribbon-like filaments of the SN 1006 remnant seem to float in space. *Hubble*'s high resolution image provides a close-up view of a 5 light-year segment of the structure, less than a third of its overall length.

The remnant's expansion has now slowed to only 6 million miles an hour, or 100,000 miles a minute. Even at this tremendous speed, it takes years between observations to detect any motion of the ribbon filament against the background stars because it is so distant—approximately 7,000 light-years.

The stellar explosion was seen in 1006 A.D. by observers from Africa to the Far East, although in reality, the star had detonated 7,000 years before—the travel time for the event's light to reach Earth. It was probably the brightest "star" ever observed by humans. Over the course of a few days, the star became brighter than the planet Venus. For several weeks, people could see it with unaided eyes—even during the day—and it remained visible in dark skies for approximately two-and-a-half years before fading away. It is likely that the supernova's progenitor star was a white dwarf, the burned-out relic of an ordinary star. The dead star had probably become unstable after siphoning too much matter off an orbiting companion star.

This image is a composite of visible (or optical), radio, and x-ray data of the full shell of the supernova remnant from SN 1006. Only a small linear filament in the northwest corner of the shell is visible in the optical data. The shell has an angular size of roughly 30 arcminutes (0.5 degree, or about the size of the full Moon). The small orange box along the bright filament at the top of the image corresponds to the dimensions of the *Hubble* image seen on page 82.

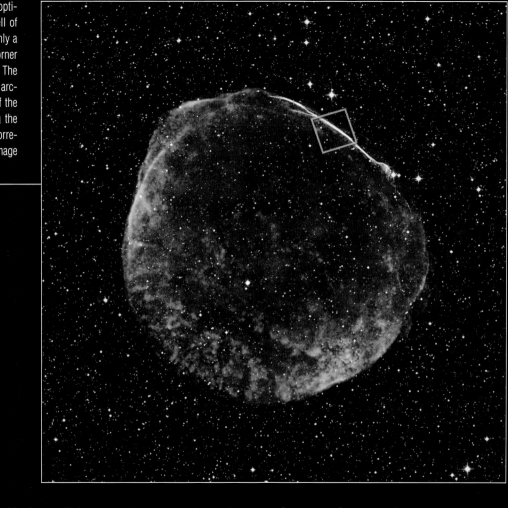

The *Hubble* image is a composite of hydrogen-light observations taken with the Advanced Camera for Surveys in February 2006, and the Wide Field Planetary Camera 2 in April 2008. The supernova remnant was assigned a red hue in the image to correspond to a prominent spectral line of hydrogen, which lies in the red portion of the electromagnetic spectrum. The orange-colored dots and smudges in the image are distant background galaxies, while the white points of light are stars in

An Edge-on View of the Thin, Expanding Shell from a Supernova

SN 1006: Example of a full-shell supernova remnant

Surface section of expanding bubble

Thin ghostly wisps of matter seen in cross-section of the expanding bubble's irregular surface

The expanding shell of a supernova remnant is like a translucent bubble of gas with surface ripples and disturbances caused by varying gas densities. The structure seen by *Hubble* corresponds to a part of space where the shock wave is sweeping up tenuous hydrogen gas, heating it, and causing it to radiate in visible light. When viewed from the side, this flattened sheet of heated gas looks like a delicate, twisting ribbon.

Further Reading

Blair, W.P., et al., "*Hubble Space Telescope* Imaging of the Primary Shock Front in the Cygnus Loop Supernova Remnant," *The Astronomical Journal*, **129**, 2268–2280, 2005.

Goldstein, B.R., and H.P. Yoke, "The 1006 Supernova in Far Eastern Sources," *The Astronomical Journal*, **70**, 748–753, 1965.

Raymond, J., et al., "The Preshock Gas of SN 1006 from Hubble Space Telescope Advanced Camera for Surveys Observations," *The Astrophysical Journal*, **659**, 1257–1264, 2007.

Reddy, F., "Supernova Aftermath: A Supernova's Impact Lives on Long Beyond Its Fading Light," *Astronomy*, **34**(6), 42–43, 2006.

Semeniuk, I., "Blasts from the Past: The Dazzling Light of Long-gone Supernovae Is Still Visible If You Know Where to Look," *New Scientist*, **194**, 46–50, 2007.

van den Bergh, S., "The Optical Remnant of the Lupus Supernova of 1006," *The Astrophysical Journal*, **208**, L17, 1976.

Winkler, P.F., "SN 1006: A Thousand-year Perspective," *Highlights of Astronomy*, **14**, 301–302, 2007.

William P. Blair is an astrophysicist and research professor in the Department of Physics and Astronomy at The Johns Hopkins University. His main scientific interests lie in the areas of gaseous nebulas and the interstellar medium, but his particular focus is supernova remnants. He has used *Hubble* to observe various supernova remnants, both in the Milky Way and in nearby galaxies, and he was instrumental in the study of the SN 1006 remnant. He has been Chief of Operations for the *Far Ultraviolet Spectroscopic Explorer* (FUSE) project at Johns Hopkins since 2000. Prior to FUSE, he worked for many years on the *Hopkins Ultraviolet Telescope*, which flew twice on the Space Shuttle. He was born in Garden City, Michigan, a suburb of Detroit. Dr. Blair earned a B.A. in mathematics and physics from Olivet College in Olivet, Michigan. He obtained an M.S. and Ph.D. in astronomy from the University of Michigan at Ann Arbor in 1981 and spent three years at the Harvard-Smithsonian Center for Astrophysics prior to coming to Hopkins. (Photo credit: W. Kirk, the Johns Hopkins University)

Probing the Atmospheres of Exoplanets

When the *Hubble Space Telescope* was launched in 1990, planets around other stars had not yet been discovered. Since 1995, however, when the first extrasolar planet, or "exoplanet," was detected, the rate of discovery of new exoplanets and external solar systems has been truly remarkable. (Note: both "exoplanet" and "extrasolar planet" are used interchangeably in this article.) By early 2009, more than 340 planets had been found orbiting other stars—almost all of them discovered indirectly by ground-based telescopes as a result of the small, but measurable, gravitational "wobble" they impose on the stars they circle.

In the first half-decade following the initial discoveries, however, astronomers were able to derive very few characteristics about these worlds beyond their masses and basic orbital characteristics. Although it was a very important and necessary beginning to know masses and orbits, the information about exoplanets, which is most crucial to researchers exploring the possibility of life elsewhere in the universe, concerns the nature of their atmospheres. What atoms and molecules are present in an exoplanetary atmosphere and in what relative fractions? What are the temperatures and pressures? Is there evidence for cloud structure and wind patterns? Finally, when the data pertaining to all of these questions are considered, is the exoplanet's atmosphere potentially supportive of life?

As difficult as the first discovery of exoplanets was, the detection and characterization of their atmospheres is significantly more challenging and represents a true giant step forward toward the ultimate goal: the discovery of exoplanets on which life is likely to exist. This is a journey we are just beginning, and although *Hubble* is not likely to realize the full dream, it has undeniably led the way in the exciting new field of finding and measuring exoplanetary atmospheres. At no time did the telescope's designers in the 1970s and 1980s think that such possibilities existed. This is but one example among many of how the most exciting science to emerge from *Hubble* is often that which is unanticipated.

This artist's concept shows exoplanet HD 189733b orbiting its parent star. *Hubble* detected the organic molecules methane, carbon dioxide, carbon monoxide, and the inorganic molecule water, in its atmosphere. The Jupiter-sized exoplanet orbits a star 63 light-years away in the constellation Vulpecula.

This artist's illustration shows extrasolar planet HD 209458b orbiting very close to its host star. The planet is about the size of Jupiter. Unlike Jupiter, the planet is so hot that its atmosphere is "puffed up." Starlight is heating the planet's atmosphere, causing hot gas to escape into space, like steam rising from a boiler.

Exoplanet HD 209458b

The exoplanet HD 209458b is what is known as a "hot Jupiter"—a class of planet about the size of the gas giant Jupiter, but orbiting closer to its parent star than the tiny innermost planet Mercury circles the Sun. The planet's orbital period is only 3.5 days. The star, HD 209458, is a yellow, Sun-like star, visible with binoculars in the autumn constellation Pegasus, and lying approximately 150 light-years from Earth.

The most important aspect of the exoplanet's orbit is that it is almost "edge-on" to our line of sight, which means that once per orbit it passes in front of, or "transits," the disk of the star. Similarly, it goes into "eclipse" once per orbit behind the star. This edge-on orbital geometry—which is shared by some other exoplanetary systems—offers profound advantages to the study of exoplanetary atmospheres. In 2001, *Hubble* was the first telescope ever to detect and begin the characterization of these atmospheres.

HD 209458b's atmospheric composition was probed when the planet transited the star. As the light from the parent star passed briefly through the atmosphere along the edge of the planet, the gases in the atmosphere imprinted their unique absorption signatures on the starlight.

Using this technique, Lead Investigator David Charbonneau of the Harvard-Smithsonian Center for Astrophysics used *Hubble*'s Space Telescope Imaging Spectrograph (STIS) to detect the presence of sodium in the planet's atmosphere. Because the detected sodium was three times weaker than had been predicted, the researchers were able to make some tentative deductions about the existence of clouds, which act to mask—and thereby lower—the measured level of sodium. Only the precision afforded by space-based spectroscopy allowed this extraordinarily difficult measurement to be made. As of late 2008, similar attempts to detect sodium in HD 209458b with giant, ground-based telescopes far larger than *Hubble* have been unsuccessful.

In 2003, an international team of astronomers, led by Alfred Vidal-Madjar of the Institut d'Astrophysique de Paris, Centre National de la Recherche Scientifique (CNRS), France, again studying HD 209458b with *Hubble*, discovered—in the first discovery of its kind—the atmosphere of an exoplanet evaporating off into space. The planet's outer atmosphere is evidently so heated by its parent star that it begins to escape the planet's gravity. The *Hubble* observations revealed a hot and puffed-up evaporating hydrogen atmosphere surrounding and following the planet—much like the tail of a comet.

Additional observations conducted in 2007 by astronomers Gilda Ballester, David K. Sing, and Floyd Herbert, provided additional details of this gaseous envelope. Their research revealed the temperature of the layer in HD 209458b's upper atmosphere, where the gas becomes so heated it escapes. At the top of this layer, the temperature skyrockets from about 1,340 degrees Fahrenheit to as much as 26,540 degrees Fahrenheit—hotter than the surface of the Sun. Given the high evaporation rate observed, it is theorized that much of the planet may eventually disappear, leaving behind only a dense core

This is an artist's impression of the Jupiter-sized planet HD 189733b being eclipsed by its parent star. (Figure credit: M. Kornmesser/ESA and STScI)

Exoplanet HD 189733b

Exoplanet HD 189733b was discovered orbiting the star HD 189733A in 2005, when astronomers observed the tiny drop in light from the star/planet system when the planet transited across the face of the star. Over the past few years, many additional observations have been taken of this system. Only 63 light-years away, it is the most accessible of all of the known transiting hot Jupiters. The combination of a large planet and relatively small parent star—only 76 percent of the diameter of our Sun—makes this planet comparatively easy to detect.

HD 189733b, which is slightly more massive than Jupiter, is so close to its parent star it takes just over two days to complete an orbit. At a scorching 1,700 degrees Fahrenheit, its atmosphere is about the same temperature as the melting point of silver. In 2007, a team led by Frédéric Pont from the Geneva University Observatory in Switzerland used *Hubble*'s Advanced Camera for Surveys (ACS) to observe the planet. Using a special grism (a cross between a prism and a diffraction grating) within ACS, the astronomers made extremely accurate measurements of the spectrum of HD 189733b.

They had expected to see the fingerprints of sodium, potassium, and water in the spectrum—and instead detected none. This finding, combined with the distinct shape of the planet's spectrum, pointed to the presence of high level hazes in its atmosphere, which would block the anticipated spectral lines. Estimated to be at an altitude of roughly 620 miles, hazes on HD 189733b are believed to consist of tiny particles 1,000 times smaller than the diameter of a pinhead. They may be condensates of iron, silicates, and aluminum oxide dust (the compound on Earth that makes up the mineral sapphire).

Adding to the overall knowledge of HD 189733b by the group was the finding that the transiting exoplanet's light curve did not reveal the signature of any Earth-sized moons nor any discernable Saturn-like ring system. Thus, bit by bit, the presence or absence of spectral features has begun to permit real analysis of the conditions present within planetary systems around other stars.

The Latest Big News

In 2008, *Hubble* enabled two other significant advances in the technique of analyzing the atmospheres of transiting/eclipsing exoplanets with the first-ever detection of an organic (carbon-bearing) molecule. Mark Swain of NASA's Jet Propulsion Laboratory in Pasadena, California led the team that made this discovery in the atmosphere of HD 189733b using extensive observations from *Hubble*'s Near Infrared Camera and Multi-Object Spectrometer (NICMOS) when HD 189733b was transiting the stellar disk. The team also confirmed the existence of water molecules in the planet's atmosphere, a discovery made originally by NASA's *Spitzer Space Telescope* in 2007.

Detection of organic molecules is a crucial step toward finding evidence for life outside our solar system. The molecule detected was methane, whose chemical symbol is CH_4 and which, under the right conditions, may be produced by living organisms. Some of the very first forms of life on Earth may have been methanogens, a form of bacteria that produces methane as a metabolic byproduct. HD 189733b is too close to its star and too hot to support life as we know it. The discovery demonstrates, however, that astronomers can detect organic molecules and water in the atmospheres of distant transiting planets.

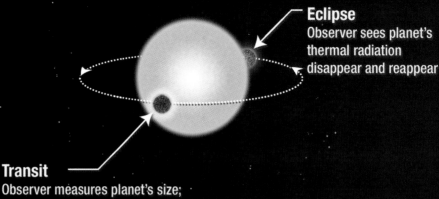

Different types of information can be gained from watching a planet pass in front of and behind its parent star as viewed from Earth.

Eclipse
Observer sees planet's thermal radiation disappear and reappear

Transit
Observer measures planet's size; sees star's radiation transmitted through planet's atmosphere

Transits, however, are only half of the story. When HD 189733b goes into "eclipse"—that is, behind its star—astronomers are given another chance to characterize its exoplanetary atmosphere. This method looks for "what disappeared" from the combined star–planet spectrum taken before the eclipse. Spectral signatures that disappear in this way are attributed to the exoplanet itself. The reason is that the subtraction of the pure star spectrum from the combined star/planet spectrum yields the spectrum of just the exoplanet. Using NICMOS and the eclipse method, Swain and his colleagues added to the inventory of water and methane for HD 189733b by confirming carbon monoxide (CO, suggested earlier to be present by Charbonneau and colleagues) and detecting carbon dioxide (CO_2) outright. Altogether, that makes for a chemical inventory detected, to date, of water and three organic molecules in one exoplanetary atmosphere. Clearly, the detailed spectroscopic measurement of exoplanets in edge-on orbits is capable of returning extremely important information—a reality that was simply unimaginable in the 1990s.

The discovery of methane on HD 189733b is, however, not without mystery. In our solar system, methane is not seen in the hot atmosphere of Venus, but only in those planetary atmospheres much colder than Earth's. Most astronomers expected a hot planet such as this one to produce carbon monoxide, not methane. Scientists are still puzzling over its source. One explanation may be that the atmosphere in HD 189733b's night sky is cold enough to produce methane, which is then transported by fast winds to the day/night edge—the region probed in this study. Or perhaps the methane results from a chemical reaction triggered by the light from the parent star. Or even more simply, maybe it is outgassed from inside.

Transiting (and Eclipsed) Exoplanets

Astronomers cannot physically sample celestial objects in a laboratory the way chemists or biologists can sample the subjects of their study. Most of what we know about celestial objects comes from the light they emit, reflect, or absorb. One of the most important tools used by astronomers is spectroscopy, in which the light is broken into its component wavelengths.

When an exoplanet passes in front of its star as detected from Earth, the planet's atmosphere absorbs more of the starlight at wavelengths where the atmosphere is opaque and less at wavelengths where the atmosphere is transparent. Mark Swain and his team used *Hubble*'s near-infrared NICMOS camera to construct an infrared spectrum of exoplanet HD 189733b, as seen in transmitted starlight. The planet's atmosphere soaked up infrared light in a pattern expected of methane and water molecules. It produced a spectrum that shows the distinctive absorption features of methane and water vapor in the planet's atmosphere.

Another type of spectrum, called an emission spectrum, was taken as HD 189733b passed behind its companion star. The occultation of this planet allowed the opportunity to subtract the light of the star alone—when the planet is blocked—from that of the star and planet together prior to eclipse. This allowed the isolation of the emission of the planet alone, making possible a chemical analysis of its "dayside" atmosphere. In this analysis, carbon monoxide and carbon dioxide were detected.

As starlight passes through the exoplanet's atmosphere, methane absorbs some of the light and leaves its distinctive "fingerprint."

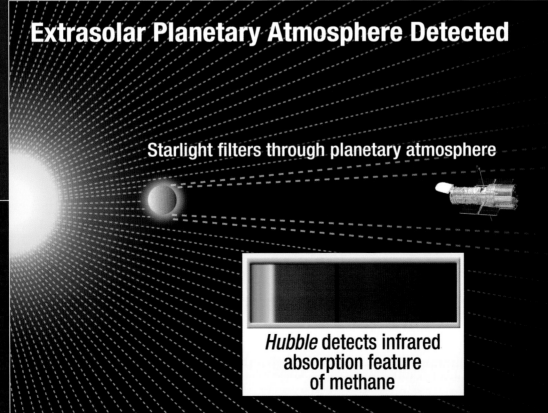

Looking Toward the Future

Hubble's observations of transiting planets are proof in concept that large, infrared-optical space telescopes can use transits and eclipses to effectively yield information about exoplanet atmospheres. With the original STIS observation of HD 209458b and the NICMOS and ACS observations of HD 189733b, three separate *Hubble* instruments have now shown their ability to contribute data in this challenging arena. Assuming the successful astronaut servicing of *Hubble* in 2009, there should be additional contributions by the observatory's two new instruments, the Cosmic Origins Spectrograph (COS) and Wide Field Camera 3 (WFC3), which also has a near-infrared grism. With these instruments operating, the number of molecules discovered is expected to grow rapidly.

Eventually, with the larger, more sensitive *James Webb Space Telescope*, astronomers may be able to perform molecular spectroscopy at infrared wavelengths of the exoplanets that lie in a star's habitable zone. This is also a major objective for future, very large aperture ultraviolet-optical telescopes in space.

Such next-generation space telescopes will enable the possibility of actually identifying biomarkers—decisive spectral signatures that would establish that biological activity is present. These biomarkers include oxygen, ozone, methane, nitrous oxide, and chlorine. To find them, scientists would look for molecules in the exoplanet's atmosphere that are present in amounts not explainable through thermochemistry or photochemistry, but require some other mechanism for sustaining unusual abundances.

Hubble's observations of far-flung planetary systems have given us important first data in the quest to answer two grand questions: How unique is Earth, and how common is life in the universe? Using *Hubble*, *Spitzer*, *Webb*, and other planned telescopes, scientists should soon have all the necessary tools to provide us more definitive answers.

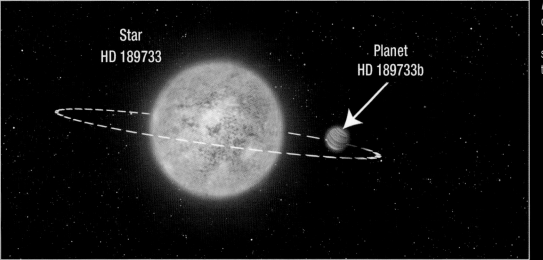

Hubble detected carbon monoxide and carbon dioxide in the atmosphere of exoplanet HD 189733b by taking the combined spectra of the star and planet, then subtracting the star's spectrum from the combined spectra.

Recognizing Life

In their search for life on other planets, scientists have only one blueprint: the mix of elements such as methane, oxygen, and ozone that make up the atmosphere of Earth, a planet that orbits a comfortable distance from its star, the Sun. There could be, however, other recipes for life. Scientists may not even recognize life if they find it.

Even on Earth, salty lakes, polar ice caps, volcanic vents, and other places deemed inhospitable to life were found just a few decades ago to have colonies of hardy microbes. One way to study distant worlds for life is to make models of the environments of those possible worlds in the lab. Researchers at the Virtual Planetary Laboratory at the California Institute of Technology's Spitzer Science Center suggest that not all life will metabolize food and energy exactly like life on Earth. So, they are developing models of planets that generate alternative environments that might bear the signature of life.

One computer model they produced puts Earth around different types of stars. They placed our planet, for example, around a red dwarf—a common type of star that is not as hot as the Sun. The model predicted a decrease in the amount of ozone, but an increase in methane—both biosignatures gases. The researchers thus discovered that many of the biosignature gases in the atmospheres of planets around different stars survive longer in their atmospheres and are much easier to see.

The environments of complex planets are also expected to change over time. Earth is one example. Earth's atmosphere has dramatically changed over the last 4.6 billion years, driven largely by the actions of life.

About 2.3 billion years ago, there was a rapid rise of oxygen in Earth's atmosphere resulting from colonies of blue-green algae that produced oxygen as a waste byproduct. This eventually allowed multicellular life to develop by providing a new chemical source of energy, and by building the ozone layer to block our ultraviolet light that is destructive to organic compounds.

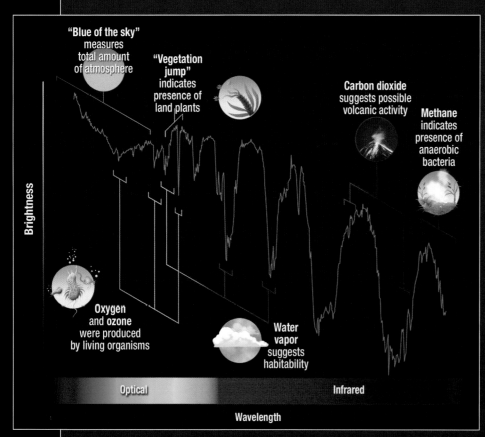

Earth's spectroscopic signatures indicate the possible conditions conducive to life.

Further Reading

Barman, T., "Identification of Absorption Features in an Extrasolar Planet Atmosphere," *The Astrophysical Journal Letters*, **661**, L191–L194, 2007.

Charbonneau, D., et al., "Detection of an Extrasolar Planet Atmosphere," *Astrophysical Journal*, **568**, 377–384, 2002.

Deming, D., "Astrophysics: Quest for a Habitable World," *Nature*, **456**(7223), 714–715, 2008.

Siegfried, T., "Exoplanet Jackpot Shows Astronomers Are Looking for Worlds in All the Right Places," *Science*, **316**(5830), 1420–1421, 2007.

Udry, S., and N.C. Santos, "Statistical Properties of Exoplanets," *Annual Review of Astronomy & Astrophysics*, **45**(1), 397–439, 2007.

Mark Swain leads an effort to identify molecules in exoplanet atmospheres using infrared spectroscopy. Results from this effort include the first detection of methane and carbon dioxide in an exoplanet atmosphere. Born in Boulder, Colorado, Swain received a B.A. in physics from the University of Virginia in 1989, and a Ph.D. in physics and astronomy from the University of Rochester in 1996. He is currently a research scientist at the Jet Propulsion Laboratory. His main scientific focus is using molecular spectroscopy to probe the conditions, composition, and chemistry of exoplanet atmospheres.

Magnetic Filaments in an Active Galaxy

NGC 1275 is located about 235 million light-years away in the constellation Perseus. Also known as Perseus A, the galaxy lies at the heart of the Perseus cluster, a rich collection of more than 500 galaxies. The largest galaxy in the cluster, NGC 1275 is also one of the closest giant elliptical galaxies.

The galaxy is home to a supermassive black hole that is accreting material at a very rapid rate. The presence of this central black hole makes the giant galaxy a well-known radio source and a strong emitter of x-rays. NGC 1275 is classified as an active galaxy—one that shows physical activity near its core and radiates at higher-than-average luminosity levels over some or all of the electromagnetic spectrum. Gas, swirling near its black hole, is causing energetic activity that creates spherically-shaped "bubbles" of material to be ejected into the surrounding galaxy cluster. Extremely long filaments of cold gas are seen emanating from the core of the galaxy and extending out in the wake of the rising "bubbles."

These gossamer-appearing filaments have withstood the hostile, high-energy environment of the galaxy cluster for more than 100 million years. The gravitational forces alone within NGC 1275 would destroy the filaments within 10 million years if not for additional forces keeping them in equilibrium. The filaments are also likely experiencing turbulence from the rising hot "bubbles."

Indeed, the long, gaseous tentacles stretch out beyond the galaxy into the multimillion degree, x-ray–emitting gas that fills the cluster. The tentacles provide evidence in visible light of the intricate relationship between the central black hole and the surrounding gas in the galaxy cluster. But why have they not heated up, dispersed, and evaporated by now, or simply collapsed under their own gravity?

Fine, thread-like filamentary structures surround active galaxy NGC 1275. The red filaments are composed of cool gas suspended by a magnetic field, and are enveloped by the 100-million-degree Fahrenheit hot gas in the center of the Perseus galaxy cluster.

Magnetic Filaments in Erupting Galaxy

The top x-ray image provides the larger context for the detail of the *Hubble* image, below. The pink outline overlying the *Hubble* detail indicates a "bubble" formed by the magnetic forces emanating from the center of NGC 1275.

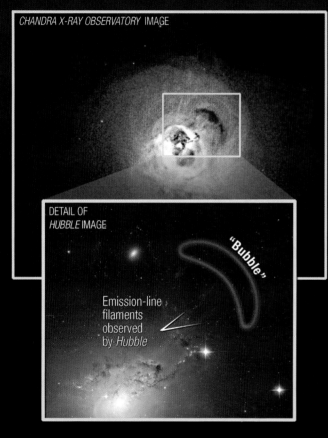

As illustrated in these *Chandra* and *Hubble* images, gas swirls near NGC 1275's black hole, creating energetic activity that ejects "bubbles" of material into the surrounding galaxy cluster. Immensely long filaments form when cold gas from the galaxy's core is dragged out in the wake of the rising "bubbles."

Using *Hubble* data, a team of astronomers led by Andy Fabian from the University of Cambridge, UK, has proposed a solution—magnetic fields hold the charged gas in place and resist the forces that would distort their filamentary structure. The team deduced this by resolving, for the first time, the individual threads of ionized gas that make up the filaments. The team found that the tentacles are only about 200 light-years wide, are often very straight, and extend for up to 20,000 light-years. Such thin filaments require constraining magnetic fields for integrity and survival. The magnetic field lines provide a skeleton that holds the filaments together against their surrounding forces.

The thinner the filament, the stronger the maintaining magnetic field must be. *Hubble* data enabled the team to measure individual filaments and deduce the strength of the magnetic fields that are in equilibrium with the hot gas. The astronomers determined that the strength of the fields are only about 1/10,000th that of Earth's, but because they extend over galaxy-sized regions, they contain an immense amount of magnetic energy.

The team also found the amount of gas contained in a typical thread is around one million times the mass of our own Sun. The gas forming them is also roughly the same mix of hydrogen, helium, and other elements that comprise the Sun.

Most ionized gases possess magnetic fields. Churning gas can wind up the fields, making them stronger. The black hole, while not the source of the magnetic field, causes motions that stir up the gas and amplify the fields.

The filamentary system in NGC 1275 provides a striking visual example of the workings of extragalactic magnetic fields. These structures may be common on much smaller scales in normal galaxies. In our own Milky Way galaxy, there is an arc-like feature near the central black hole that is believed to be hot plasma flowing along magnetic field lines.

Immense networks of filaments similar to those in NGC 1275 are found around many other, more remote central cluster galaxies, but they cannot yet be observed with comparable resolution. For now, the team will apply their understanding of NGC 1275 to interpret observations of these more distant galaxies.

Further Reading

Conselice, C.J., J.S. Gallagher III, and R.F.G. Wyse, "On the Nature of the NGC 1275 System," *Astronomical Journal*, **122**, 2281–2300, 2001.

Fabian, A., R. Johnstone, J. Sanders, et al., "Magnetic Support of the Optical Emission Line Filaments in NGC 1275," *Nature*, **454**, 968–970, 2008.

Hatch, N.A., C. Crawford, and A. Fabian, "Detections of Molecular Hydrogen in the Outer Filaments of NGC1275," *Monthly Notices of the Royal Astronomical Society*, **358**, 765–773, 2005.

Kent, S.M., and W.L.W. Sargent, "Ionization and Excitation Mechanisms in the Filaments Around NGC 1275," *Astrophysical Journal*, **230**, 667–680, 1979.

Salome, P., "Cold Gas in the Perseus Cluster Core: Excitation of Molecular Gas in Filaments," *Astronomy & Astrophysics*, **484**, 317–325, 2008.

One of the United Kingdom's foremost high-energy astrophysicists, Andy Fabian has built his career by using the techniques of x-ray astronomy to investigate extreme astrophysical conditions. His interest in space astronomy dates to his childhood in the U.K. He completed his undergraduate work in physics at King's College, London, and earned his Ph.D. from the Mullard Space Science Laboratory at the University College London. Dr. Fabian is a Royal Society Professor at the Institute of Astronomy at the University of Cambridge and President of the Royal Astronomical Society. His research interests are in accreting black holes and clusters of galaxies.

Interacting Galaxies

Far from being solitary and isolated island universes, many galaxies are found to be interacting. Their close encounters can lead to spectacular mergers and spawn vast amounts of new star formation. Astronomers estimate that in the nearby universe, 1 out of every 20 gas-rich disk galaxies, like our Milky Way galaxy, is in the act of colliding. Galaxy mergers were much more common in the past, however, when the expanding universe was smaller. For example, in a class of galaxies called ultraluminous infrared galaxies, which have quasar-like luminosities and account locally for one out of every million massive galaxies observed, collisions were 100 times greater when the universe was half of its current age. This is largely because the galaxies were much closer to each other and very gas-rich.

Fifty-nine new *Hubble* optical images of interacting galaxies were released in 2008. Most of the images are part of a large investigation of luminous and ultraluminous infrared galaxies called the Great Observatory All-sky (Luminous Infrared Galaxies) LIRG Survey (GOALS) project—a survey that combines observations from *Hubble*, the NASA *Spitzer Space Telescope*, the NASA *Chandra X-Ray Observatory*, and the NASA *Galaxy Evolution Explorer*. The *Hubble* optical observations were led by Professor Aaron S. Evans from the University of Virginia and the National Radio Astronomy Observatory (USA). *Hubble*'s specific role in the project was to employ its high optical resolution to study star formation and active galactic nuclei processes—the latter of which is the feeding and build-up of supermassive black holes in the centers in these galaxies— and the possible connection between these phenomena.

Through the use of computer modeling and detailed *Hubble* imagery, scientists have studied how gravity choreographs the motions and the changing morphological shapes of colliding galaxies. These epic clashes occur at a glacial pace—taking about a billion years to complete. Each image of a galactic encounter is at best only a "freeze frame" in an unimaginably long movie. By examining a large sample of galaxies at different stages in the collision process, however, astronomers hope to learn how star formation, black hole activity, and galaxy evolution characteristically progress.

This image is one in a collection of 59 images of merging galaxies taken by *Hubble* and released on the occasion of its 18th anniversary, April 24, 2008. It is known to astronomers as Arp 240. Both galaxies likely harbor supermassive black holes in their centers and are actively forming new stars in their disks. Arp 240 is located in the constellation Virgo, approximately 300 million light-years away. With the exception of a few foreground stars, all the objects in this image are galaxies.

While galaxies collide, with very rare exceptions, the stars within them do not. This is because so much of a galaxy is simply empty space, with distances between stars about 100 million times larger than their stellar diameters. What collides is the gas and dust between the stars, which produces a torrent of new star formation. Our own Milky Way galaxy contains the debris of the many smaller galaxies it has encountered and devoured in the past—like the Sagittarius Dwarf Elliptical Galaxy it is currently absorbing. Several billion years from now, the Milky Way galaxy will collide with the Andromeda galaxy, our closest large galactic neighbor. A near twin to the Milky Way, Andromeda is now headed toward us at about 670,000 miles per hour. *Hubble*'s view of galactic collisions gives scientists an idea of what the Milky Way will experience when its inevitable merger with Andromeda occurs.

Although the visual results of galactic interactions are very different depending on what types of galaxies are involved, from what angles they approach, and on how they collide, there is one constant agent at work: gravitationally-produced tidal disruption. As the gravitational fields linking the stars and star clusters in each galaxy begin to interact, strong tidal effects distort the original patterns. This leads to new structures, and eventually—over long periods of time—to a stable new configuration.

The pull of the Moon that produces the twice-daily rise and fall of Earth's oceans illustrates the nature of tidal interactions. Tides between galaxies, however, are much more disruptive than oceanic tides for two main reasons. First, stars in galaxies, unlike the matter that comprises Earth, are bound together only by gravity, so they have no other forces holding them together. Second, galaxies can pass much closer to each other, relative to their size, than do Earth and the Moon. The billions of stars in each interacting galaxy move individually, following the pull of gravity from all the other stars, so the tidal forces can produce intricate and varied effects as the galaxies pass by and through one another.

The first sign of an interaction is a bridge of matter between the two galaxies as gravity teases out dust and gas from the approaching pair. As the outer reaches of the galaxies begin to intermingle, streamers of gas and dust, known as tidal tails, stretch out and sweep back to wrap around the cores. These long, often spectacular, tidal tails can persist long after the main collision is over.

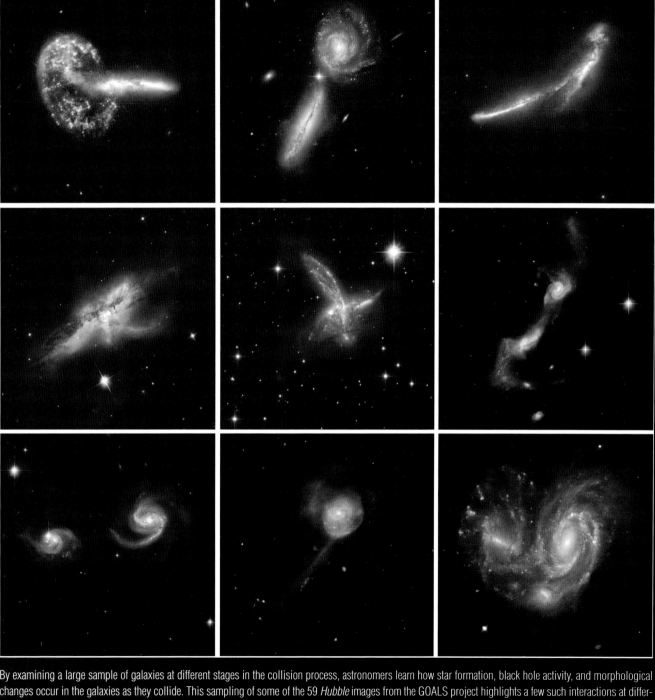

By examining a large sample of galaxies at different stages in the collision process, astronomers learn how star formation, black hole activity, and morphological changes occur in the galaxies as they collide. This sampling of some of the 59 *Hubble* images from the GOALS project highlights a few such interactions at different stages of their collision/merger process. Astronomers estimate that 1 out of every 20 disk galaxies (like the Milky Way) in the nearby universe is in the act of colliding.

Three Examples of Interacting Galaxies

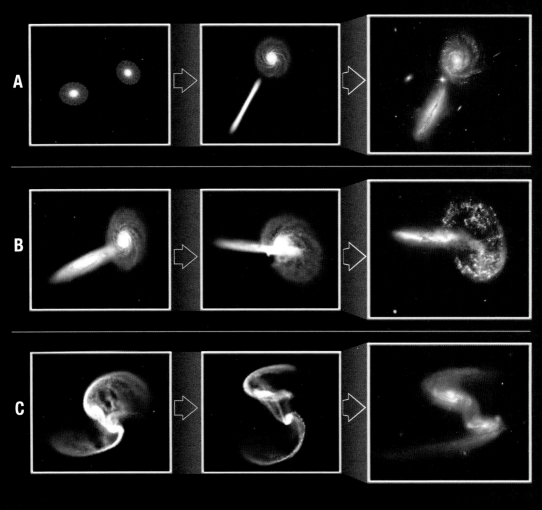

Galaxy collisions, and any subsequent mergers, can be modeled based on astronomical images of galaxies in different stages of that process and on advancements made in the related computer-based mathematical algorithms. This illustration depicts three images taken by *Hubble* of interacting galaxies (right-most panels), as well as two panels each of artist's concepts of the interactions leading up to the "snapshots" captured by *Hubble*.

As the galaxy cores approach each other, their gas and dust clouds are buffeted and accelerated dramatically by the conflicting pull of matter from all directions. These forces can result in shockwaves rippling through the interstellar clouds. Gas and dust can then be siphoned into the active central regions, either fueling bursts of star formation that appear as characteristic blue knots of young stars, or feeding the supermasssive black holes that reside at the centers and creating superluminous quasars. As the clouds of dust build throughout the collision zone, they are heated and make their parent galaxies the brightest infrared objects in the sky—emitting up to several trillion times the luminosity of our Sun, but radiated as infrared energy, not as visible light.

There were concerns, in fact, that using *Hubble* within the GOALS project to study star formation in visible light in these luminous infrared galaxies would not be very effective. These galaxies become "infrared" luminous by producing new stars and feeding their giant black holes in dust-enshrouded regions of the galaxy that are opaque to visible light. Much to the contrary, this portion of the survey has been a major success. The *Hubble* data have revealed evidence of star formation in most of the infrared galaxies, and not only in the extended disks and tails, but also in the central regions of the galaxies where one would expect the star formation to be completely enshrouded by dust (based on the *Spitzer Space Telescope* imaging survey). These star clusters in the central regions that are bright in

Interacting galaxies can spawn large areas of new star formation—revealed as bright knots of clustered blue stars. This *Hubble* image of interacting spiral galaxies, NGC 6050 and IC 1179, captures many such regions. Located 450 million light-years away, the galaxies are part of the large Hercules Galaxy Cluster. The small galaxy at the top and between the two spirals also appears to be involved in the merger. Follow-up investigation with spectroscopy is needed to confirm this association.

visible light are essentially tracers of the more embedded star formation that accounts for the bulk of the energy generated in these galaxies.

Interacting galaxies are not only visually spectacular, but also scientifically useful encounters. Their study by observatories operating across the electromagnetic spectrum enlarges our knowledge of the complex processes that govern the formation of stars, black holes, and ultimately, the galaxies themselves. *Hubble* observations are a key part of this ongoing study.

Arp 81 is a strongly interacting pair of galaxies, seen about 100 million years after their closest approach. The encounter has pulled a long tail out of one of the galaxies that has now wrapped behind the pair. The collision has triggered extensive star formation between the two galaxies, revealed in its clusters of bright blue stars.

Further Reading

Benchich, E.A., "What Happens When Galaxies Collide?" *Astronomy*, **33**(9), 32–37, 2005.

Cox, T.J., and A. Loeb, "The Collision Between the Milky Way and Andromeda," *Monthly Notices of the Royal Astronomical Society*, **386**, 461–474, 2007.

Kitchin, C.R., *Galaxies in Turmoil*. New York: Springer, 2007.

Evans, A.S., et al., "Off-Nuclear Star Formation and Obscured Activity in the Luminous Infrared Galaxy NGC 2623," *The Astrophysical Journal*, **675**(2), L69–L72, 2008.

Schweizer, F., "Galaxy-scale Mergers and Globular Clusters," *Science*, **287**(5457), 1410–1411, 2000.

Aaron Evans is leading the effort to study the ultraviolet and optical properties of a large sample of high-luminosity infrared galaxy mergers with *Hubble*. Born in Wichita Falls, Texas, Evans spent a large fraction of his early years living in Asia. He received his B.S. degree from the University of Michigan in 1990, and his Ph.D. from the University of Hawaii in 1996. He held a postdoctoral scholar appointment at Caltech from 1996 to 1999 and a faculty appointment at Stony Brook University from 1999 to 2008 before starting a joint faculty/staff appointment at the University of Virginia and the National Radio Astronomy Observatory. He is a member of GOALS, a multi-wavelength program designed to assess the nature of infrared luminous galaxy mergers by assessing the properties of their extended stellar light and by probing their central energy sources (i.e., central starbursts and accreting supermassive black holes).

Dark Matter and Galaxy Life in a Supercluster

Like lights strung on a Christmas tree, galaxies in the massive supercluster Abell 901/902 appear to hang on mysterious, invisible branches. Mathematical models describing the origin of hydrogen, helium, and other light elements during the birth of the universe, or "Big Bang," predict that ordinary matter—the protons and neutrons that make up the stars, planets, gas, dust, and us—accounts for only a small fraction of the universe. The bigger fraction consists of some unknown material—dubbed "dark matter" by scientists—which does not emit any radiation that can be detected by conventional methods, but whose gravitational influence directs and constrains the formation of the large-scale structures of the universe (see sidebar on page 116). Recent *Hubble* observations of Abell 901/902 have helped to "illuminate" this invisible dark matter and the large cosmic "web" in which it entangles "normal" matter.

The Search for Dark Matter

Astronomers, using *Hubble*'s Advanced Camera for Surveys, have produced a detailed map of the dark matter framework in Abell 901/902, and of the hundreds of individual galaxies that trace it. The map was constructed by observing the light from more than 60,000 distant, background galaxies that are far beyond the supercluster. To reach Earth, the light from these faraway galaxies traveled through the dark matter surrounding the closer supercluster. As it did so, the light was bent by the dark matter's massive gravitational field—a phenomenon known as "gravitational lensing." Astronomers used this observed, subtle distortion of the background galaxies' shapes to compute the amount of dark matter in the light's path and then to reconstruct the dark matter distribution in the supercluster.

Gravitational lensing, a direct prediction of Albert Einstein's 1916 Theory of General Relativity, comes in two forms: strong and weak. Strong lensing dramatically bends the light from distant galaxies into arc-like shapes. Weak lensing distorts the images to a much lesser degree. Einstein realized that these small distortions could not be seen from the ground given the technology of his day, and so left this aspect of his theory to be validated by future experimentalists. Now, 93 years later,

A composite image of galaxy cluster Abell 901/902 taken with *Hubble* and the MPG/ESO 2.2-meter telescope in Chile. The magenta-tinged clumps indicate the location of dark matter as derived through analysis of an effect called "gravitational lensing," which slightly distorts the galaxies' shapes.

Hubble and the newer class of astronomical instruments have sufficient angular resolution to record even the perturbations of weak lensing. In this case, *Hubble* data were used to analyze large numbers of galaxies to find consistent, small, gravity-produced distortions.

Astronomers identified four main areas in the supercluster where dark matter has collected into dense clumps. These total 100 trillion times the Sun's mass. The areas are also the location of hundreds of old galaxies that lived through a violent history during their travels from the edges of the supercluster into the denser central region.

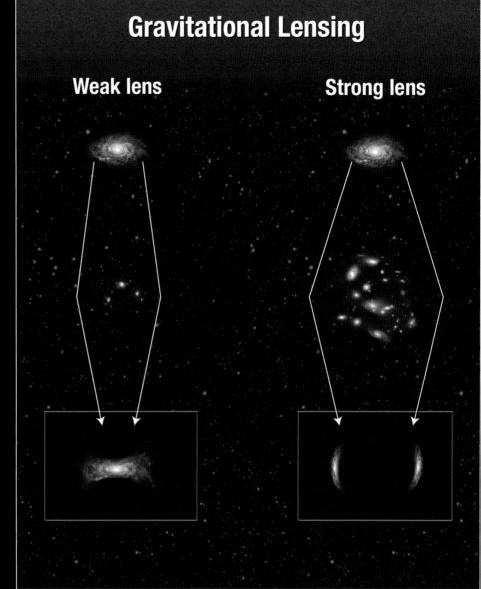

As predicted by Albert Einstein's General Theory of Relativity, a gravitational lens is formed when the light from a very distant, bright source is bent around a massive object (such as a cluster of galaxies) between the source object and the observer. With strong lensing, there are easily visible distortions such as the formation of rings, arcs, and multiple images. With weak lensing, the distortions of background sources are much smaller and can only be detected by analyzing large numbers of sources. The bowtie-shaped distortion illustrated in the weak lensing case above is an exaggerated example of what would actually be seen. Researchers Gray and Heymans used weak lensing, employing the subtle distortion of the galaxies' perceived shapes to reconstruct the distribution of intervening mass along *Hubble*'s line of sight.

The *Hubble* study revealed the distribution of dark matter in the supercluster Abell 901/902, composed of hundreds of galaxies. The magenta-tinted clumps represent a map of the dark matter in the cluster, and the supercluster galaxies lie within the clumps of dark matter. The four flanking images are details of the central images. Astronomers assembled these photos by combining a visible-light image of the supercluster taken with the MPG/ESO 2.2-meter telescope in La Silla, Chile, with a dark matter map derived from observations with the *Hubble Space Telescope*.

The dark matter map is 2.5 times sharper than a previous ground-based survey of the supercluster and shows details and nuances not possible with ground-based telescopes, whose images are further distorted by Earth's atmosphere, thereby complicating the analysis. This marks the first time that irregular clumps of dark matter have been detected and cataloged in Abell 901/902.

Dark Matter and the Cosmic Web

Dark matter is an exotic, invisible form of matter that accounts for most of the universe's mass. This mysterious matter can only be detected by its gravitational pull. It should not be confused with dark energy, a repulsive force (whose origin is currently unknown) opposing the force of gravity. Studying dark matter may eventually unlock the secret to dark energy, which influences how dark matter condenses.

In this artistic depiction of the cosmic web—the large-scale structure of the universe—each bright knot is an entire galaxy, while the purple filaments show where dark matter exists between the galaxies. The cosmic web is believed to be the skeleton of the universe. This web of dark matter formed in the very early universe because of extremely small-scale fluctuations in the density shortly after the Big Bang. The very rapid inflationary growth, which the universe underwent shortly after the Big Bang, grew these tiny fluctuations into the large-scale web-like structure we see today.

Normal matter is gravitationally attracted to the strands and clumps in the dark matter web. Its resulting contraction gives rise to the formation of stars and galaxies.

The cosmic web is still in the process of evolving as the gravity of the dark matter pulls normal matter into large clusters and groups of galaxies. At the same time, the mysterious force called dark energy is causing the expansion of our universe to accelerate. Its effect is opposite the gravitational pull of the dark matter. Dark matter influences structures to collapse and form, pulling galaxies into large cluster groups; but dark energy causes accelerating expansion, pulling structures apart. (Figure credit: Visualization by F. Summers, STScI. Simulation by L. Hernquist, Harvard University and M. White, University of California at Berkeley)

Galaxy Details

Mapping the underlying dark matter in the supercluster was just one use for the *Hubble* data, however. Astronomers also studied in detail the galaxies themselves to understand how galaxies are influenced by the environment in which they live. As the universe evolves, galaxies are continually drawn into larger and larger groups, clusters, and superclusters by the pull of gravity. On the fringes of a supercluster, galaxies are still traveling relatively slowly and feeling the first effects of the cluster environment. Galaxies located in this relative isolation appear very different from those found in the most crowded regions of a supercluster. Galaxies in the center of the supercluster are generally rounder, tending to be elliptical rather than spiral. They also tend to be full of old, red stars rather than still forming hot, young blue stars. The researchers believe environment plays a large role in this difference.

The *Hubble* survey data revealed that more collisions occur between galaxies in the regions toward which the galaxies are traveling than in the centers of the clusters. By the time the galaxies reach the cluster's center, they are moving too fast and with too much momentum to collide and merge. On the way to or from the cluster's periphery, however, they move more slowly and have more time to interact.

The Challenge Ahead

This study of dark matter and galaxies was part of the Space Telescope Abell 901/902 Galaxy Evolution Survey (STAGES), led by Meghan Gray of the University of Nottingham and Catherine Heymans of the University of Edinburgh, both in the United Kingdom. STAGES spanned one of the largest sections of sky ever observed by *Hubble*, an area requiring 80 *Hubble* images to cover the entire field. Abell 901/902 is located 2.6 billion light-years from Earth and is more than 16 million light-years across.

Having mapped the densest regions of dark matter in this supercluster, the STAGES team now wants to use *Hubble* to understand even more about this elusive substance. Even though they have made a very high resolution, detailed map, more data are required to see lower-mass filaments that they believe link together the dark matter structures and form the giant cosmic "tree." With new data, they also seek a more detailed understanding of how galaxies form, evolve, and interact with each other during their lifetimes within the supercluster environment.

Further Reading

Freeman, K., *In Search of Dark Matter*, New York, NY: Praxis Publishing Ltd., 2006.

Heymans, C., et al., "The Dark Matter Environment of the Abell 901/902 Supercluster: A Weak Lensing Analysis of the HST STAGES Survey," *Monthly Notices of the Royal Astronomical Society*, **385**, 1431–1442, 2008.

Massey, R., et al., "Dark Matter Maps Reveal Cosmic Scaffolding," *Nature*, **445**, 286–290, 2007.

Refregier, A., "Weak Gravitational Lensing by Large-scale Structure," *Annual Review of Astronomy and Astrophysics*, **41**, 645–668, 2003.

Dr. Meghan Gray was born in Halifax, Nova Scotia. While studying for a B.S. at Mount Allison University in New Brunswick, Canada, her interest in an astronomical career was inspired by summer research experiences at the Dominion Astrophysical Observatory and the Canada-France-Hawaii Telescope. In 1997, she traded in the beaches of Hawaii for the cobblestones of Cambridge to pursue a Ph.D. at the University of Cambridge. She has remained in the U.K., holding postdoctoral fellowships first at the University of Edinburgh and then at the University of Nottingham, where she is now a Science and Technology Facilities Council Advanced Research Fellow and lecturer in the School of Physics and Astronomy. Dr. Gray enjoys explaining her research on galaxy evolution to the public. She recently organized the conference "Malaysia09: Galaxy Evolution and Environment" at the University of Nottingham campus in Kuala Lumpur, Malaysia.

Dr. Catherine Heymans was born in Hertfordshire in the United Kingdom. She received a Masters in physics from the University of Edinburgh in 2000, and her Ph.D. from the University of Oxford in 2003. She has worked at the Max-Planck Institute in Heidelberg, Germany, and the University of British Columbia, Canada, and is now a senior research fellow at the University of Edinburgh. Her current work focuses on using weak gravitational lensing to understand dark matter and dark energy in the universe.

Barred Spiral Galaxies and Galactic Evolution

In a landmark study of more than 2,000 spiral galaxies, *Hubble* has found clear evidence that majestic barred spirals—galaxies that show a distinctive bar-shaped structure of stars and gas that slice across their nuclei—were far less common 7 billion years ago than they are today.

The findings come from *Hubble's* largest galaxy census, which surveyed 10 times more spiral galaxies than previous observations. The census is part of the Cosmic Evolution Survey (COSMOS), covering an area of sky nine times larger than the full Moon. This detailed look across cosmic history has provided important clues to the origin and evolution of these immense "celestial cities."

A team led by Kartik Sheth of the Spitzer Science Center at the California Institute of Technology in Pasadena discovered that only 20% of the spiral galaxies in the distant past possessed bars, compared with nearly 70% of their modern counterparts. Bars have been forming steadily over the last 7 billion years, more than tripling in number.

Galactic bars develop when stellar orbits in a spiral galaxy become unstable and deviate from a circular path. The tiny elongations in the stars' orbits grow and get locked into place, forming a bar. The bar becomes even more pronounced as it collects more and more stars in elliptical orbits. Eventually, a high fraction of the stars in the galaxy's inner region join the bar. This process has been demonstrated repeatedly with computer-based simulations.

As they develop, bars apparently become one of the most important catalysts for galaxy evolution. They force a large amount of gas towards the galactic center, fueling new star formation, building central bulges of stars, and feeding massive black holes. Galaxies are thought to initially assemble from, and grow in size through, mergers with other smaller galaxies. After this phase is complete, however, the only other dramatic way for galaxies to evolve is through the action of bars.

The magnificent barred-spiral galaxy NGC 1300 is located in the constellation Eridanus, the River. The astonishing details seen in this *Hubble* image belie the fact that the galaxy is actually 69 million light-years distant.

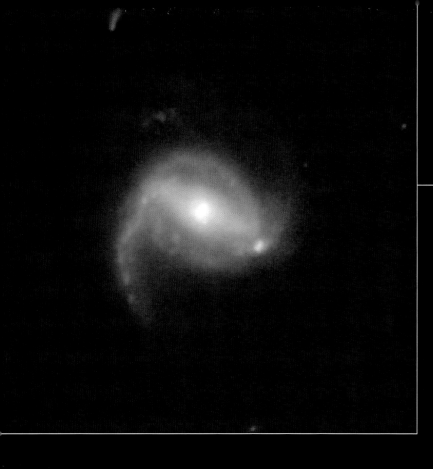

Barred spiral galaxy COSMOS 3127341 is located 2.1-billion light-years away from Earth. The galaxy is part of a landmark study of more than 2,000 spiral galaxies from the COSMOS study, the largest galaxy census conducted by the *Hubble Space Telescope*. The study's results confirm the idea that bars are a sign of galaxies reaching full maturity.

It is not clear, however, what makes stellar orbits become unstable and go on to form bars. One of the more well-accepted theories is that this happens through the interaction of stars in a galaxy with a differentially rotating disk—when a galaxy rotates not as a solid body, but with the inside rotating slightly faster than the outside. In galaxies with stars moving in a Keplerian fashion—like the planets in our solar system where the nearest to the Sun revolve the fastest—passing interactions with other stars can leave stellar orbits unstable. If there is a slight growth of structure in one part of the galaxy, it can have a dominant effect that leads to more instability, creating a situation where stars start wandering off from their smooth, circular orbits and produce a bar-like structure appearing in the overall morphology of the galaxy itself. This process will continue to be probed for a more comprehensive understanding of the various elements involved.

A second key finding from the COSMOS study is that recently forming bars are not uniformly distributed across galaxy masses. They are forming mostly in the small, low-mass galaxies, whereas among the most massive galaxies, the fraction of bars today is the same as it was in the past. This finding has important ramifications for understanding galaxy evolution. Low-mass galaxies are known to form stars at a slower pace, and *Hubble* now shows that their bars form more slowly as well.

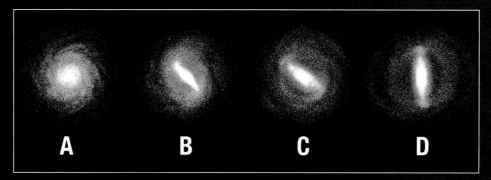

Bars are the signposts of a galaxy disk's maturity. As soon as a disk is sufficiently massive and dynamically cold, a bar forms very quickly, in only a few hundred million years. This simulation shows a typical disk galaxy at four different times in its history. Panel A shows the galaxy without a bar when the universe was only 6.2-billion years old. Panels B, C, and D show the same galaxy when the universe was 8.1-billion years old, 10.6-billion years old, and in the present-day universe at 13.7-billion years old. The bar is a long-lived and stable structure. (Figure credit: E. Athanassoula, LAM, Marseille, France)

The *Hubble Space Telescope* has proved uniquely qualified to conduct this survey. To recognize bars for a large sample of galaxies, astronomers needed panoramic images of many spiral galaxies. Ground-based telescopes with adaptive optics can recognize a bar in an individual galaxy, but it would be a very inefficient use of such telescopes to look at large samples of galaxies one by one. In contrast, a single image with *Hubble* provides a comparatively wide sample of many galaxies.

Our own Milky Way galaxy has a central bar that probably formed somewhat early, like the bars in other large galaxies in the *Hubble* survey. A better understanding of how bars formed in the most distant galaxies will eventually shed light on how such formation occurred in our home galaxy.

Further Reading

Knapen, J.H., "Barred galaxies; Stars and Bars," *Astronomy & Geophysics*, **46**, 6.28–6.33, 2005.

Knezek, P., "The Best Bar in the Neighborhood," *Sky and Telescope*, **111**(1), 32–33, 2006.

Koekemoer, A.M., and N.Z. Scoville, "The COSMOS 2-degree HST/ACS Survey, *New Astronomy Reviews*, **49**, 461–464, 2005.

Sheth, K., et al., "Evolution of the Bar Fraction in COSMOS: Quantifying the Assembly of the Hubble Sequence," *Astrophysical Journal*, **675**(2), 1141–1155, 2008.

Sparke, L., and J.S. Gallagher III, *Galaxies in the Universe: An Introduction*, New York: Cambridge University Press, 442 pp., 2000.

Kartik Sheth was born in Mumbai, India and grew up dreaming of becoming an astronaut. He was fascinated by astronomy at an early age with school field trips to the Nehru Planetarium and NASA photographs from the *Voyager* missions. He is currently a research astronomer at the *Spitzer Space Telescope* Science Center and will soon be working with the Atacama Large Millimeter Array telescope at the National Radio Astronomy Observatory. His research focuses on the formation and evolution of galaxies. As a key member of the COSMOS team and as the PI of the upcoming *Spitzer* Survey of Stellar Structure in Nearby Galaxies (S4G), he is focusing on the evolution of barred spiral galaxies. He earned a Ph.D. and M.S. in astronomy from the University of Maryland, an M.S. in physics from the University of Minnesota, and a B.A. in physics from Grinnell College.

Searching for Baryonic Matter in Intergalactic Space

Two kinds of matter exist in the universe: mysterious dark matter, and normal—or baryonic—matter. Dark matter, which neither emits nor reflects light, is an exotic form of matter that is only detected via its gravitational influence. Baryonic matter is a particle physics classification of matter made of atomic nuclei, composed of protons and neutrons. It is the ordinary matter that makes up the stars, planets, gas, dust, and us.

According to scientists studying the afterglow of the Big Bang, dark matter was about five times more abundant than baryonic matter in the early universe. Although the ratio of dark to baryonic matter should remain constant over time, observations of galaxies in the nearby, modern universe can account for only a fraction of the ordinary matter expected. The rest must be found elsewhere.

Now, in an extensive search of the local universe by *Hubble*, astronomers have definitively found about half of the missing "normal" matter in the extremely dilute gas that surrounds and stretches between the galaxies. This gas, or intergalactic medium, is believed to extend unevenly throughout all of space.

Intergalactic space might intuitively seem to be empty, but researchers have known for decades that it is home to gas that is not in condensed form like in stars and galaxies. Charles Danforth and Mike Shull of the University of Colorado have taken the most detailed census of this important component of the universe, and have shown that the intergalactic medium is in fact the reservoir for most of the normal, baryonic matter in the cosmos.

Their observations were taken along sight lines to 28 active galactic nuclei, which are the cores of distant galaxies with active black holes made extremely luminous by the gravitational energy released by matter falling into them. The intrinsically brightest active galactic nuclei are called "quasars." Danforth and Shull used their 28 active galactic nuclei—the majority of

Known as the *Hubble* Ultra Deep Field, this million-second-long *Hubble* exposure taken in 2004 shows a universe full of galaxies. These galaxies, however, contain only a fraction of the baryonic, or normal, matter in the universe. Astronomers using *Hubble* to probe the local universe have definitively found about half of the missing normal matter in the extremely dilute gas that surrounds and stretches between the galaxies.

Although Danforth and Shull's "baryon census" produced an outcome similar to *Hubble* results by Todd Tripp and colleagues in 2000, Tripp's program was based on only one quasar sightline and has thus been greatly extended and put on a much firmer statistical footing by Danforth and Shull.

Searching for dilute baryonic matter in the nearby universe was one of the key projects identified for *Hubble* when the telescope was launched in 1990. The best way to observe this local gas is through ultraviolet spectroscopy. Danforth and Shull used the ultraviolet capability of *Hubble*'s Space Telescope Imaging Spectrograph (STIS), as well as NASA's *Far Ultraviolet Spectroscopic Explorer* (FUSE), to probe the local intergalactic medium by looking at quasars. As it travels across the universe, some of the quasar light is absorbed by intervening bands of baryonic matter called "filaments." This matter leaves spectral "fingerprints" at known colors, or "wavelengths," that are characteristic of the types of atoms absorbing the light and also their line-of-sight distances. Because the Earth's atmosphere absorbs ultraviolet light, these observations cannot be accomplished with ground-based telescopes.

Taking Core Samples

To conduct a census, the researchers needed enough deep sight lines, or core samples, to have a good representation of the intergalactic medium. They used the bright light from the quasars to penetrate 650 filaments of neutral hydrogen (a bound proton and electron) in the cosmic web, and produced a limited, but tantalizing three-dimensional probe of intergalactic space. Eighty-three filaments were found laced with highly ionized oxygen, in which five electrons have been stripped away. The presence of this highly ionized oxygen (and other elements) between the galaxies is believed to trace large quantities of invisible, hot, ionized hydrogen.

How much matter to look for?

Astronomers know how much matter to look for in two independent ways, and they both have to do with phenomena that occurred in the Big Bang. Deuterium, which is a heavy form of hydrogen, was cooked up in the first three minutes of the Big Bang. It was created only in the Big Bang, and not subsequently, unlike most other elements that are formed in stars. By measuring the amount of deuterium left over today relative to ordinary hydrogen, astronomers determine a ratio that tells how many baryons were present just after the Big Bang.

The second way, which is completely independent, involves studying sound waves from the Big Bang as measured by the cosmic microwave background. Most recently, NASA's *Wilkinson Microwave Anisotropy Probe* (WMAP) performed some very detailed studies of these sound waves and confirmed that approximately 4.6 percent of all the mass/energy in the universe had to be in the form of baryons. The rest is in the form of dark matter and in dark energy, which is a mysterious, repulsive form of gravity. Of the 4.6 percent that is baryonic matter, less than 10 percent of that is locked up in galaxies. The rest must be found elsewhere.

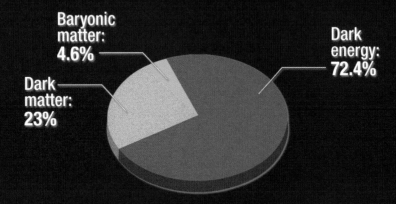

Scientists believe that a specific mix of elements resides in intergalactic space. Although astronomers can't detect the invisible, ionized hydrogen directly, oxygen in its ionized state can be detected. The oxygen accompanies the hydrogen, so scientists can deduce the existence of ionized hydrogen by finding the ionized oxygen. The oxygen "tracer" was probably created when exploding stars in galaxies spewed the oxygen back into intergalactic space, where it mixed with the pre-existing hydrogen via a shockwave that heated the oxygen to very high temperatures. These vast reservoirs of ionized hydrogen have escaped direct detection because they lack a signature in visible light and are too cool to be seen in x-rays. The oxygen has thus served as an important "proxy" for ionized hydrogen.

The 83 filaments laced with ionized oxygen also provided conclusive evidence that heavy elements from supernova explosions were blown out into very deep space, probably by galactic winds. Unlike hydrogen and helium, heavy elements were not born in the Big Bang, but in the interiors of stars. The evidence suggested that these elements have traveled several million light-years. The researchers stressed that the significance of these findings was not only in counting the baryons—it

Hubble searches for missing ordinary matter, or baryons, by looking at the light from quasars several billion light-years away. Imprinted on that light are the spectral fingerprints of the missing ordinary matter that absorbs the light at specific wavelengths (shown in the colorful spectra at right). The missing baryonic matter helps trace out the structure of intergalactic space, called the cosmic web.

was also determining how far from the galaxies they traveled and in what quantity. The team also found that about 20 percent of the baryons reside in the voids between the web-like filaments. Within these voids could be faint dwarf galaxies or wisps of matter that could turn into stars and galaxies in billions of years.

Non-baryonic dark matter is believed to collapse into filamentary structures, like huge spider webs, under the influence of gravity. Once collected, the dark matter exerts a gravitational force on the baryonic gas, pulling it into the cosmic filaments that then become visible to *Hubble*. The baryonic matter thus traces the underlying, invisible skeleton of dark matter. By uncovering the scaffolding of the universe, astronomers are also learning how galaxies formed and evolved.

This process is not very efficient. Only about 10 percent of the total baryonic matter is sufficiently condensed by gravity to form stars and galaxies. More than 90 percent was left between the galaxies. Some of this matter in intergalactic space could continue to fall into our galaxy and others; so the Milky Way still has matter falling in from intergalactic space and is still in the process of assembling.

Looking Ahead

Before *Hubble*, astronomers had found only about 10 percent of the baryonic matter in the local universe. Now, with this latest *Hubble* finding, they can account for roughly half. Taking advantage of the much greater ultraviolet sensitivity of the Cosmic Origins Spectrograph (COS), which astronauts will install on *Hubble* in 2009, scientists will be able to observe fainter, more distant objects. It is anticipated that COS could find another 10 to 20 percent of the baryonic matter in weak filaments of the cosmic web.

Probing the vast cosmic web will be a key goal for the COS, which will make possible more robust, more detailed, and significantly more numerous core samples of the cosmic web. COS will be up to 30 times more sensitive than STIS, allowing astronomers to use fainter quasars to look along many more sight lines, thus building up a more complete picture of the cosmic web, its constituents, and its physical state. Together with the larger astronomical community, the COS team hopes to observe at least 100 additional quasars and build up a survey of more than 10,000 hydrogen filaments in the cosmic web, many laced with heavy elements from early stars. After COS has completed its work, the percentage of missing baryons should be greatly reduced from what it is today.

Shull predicts that in another 10 to 15 years, astronomers will probably have to look in the x-ray range to find the rest. With much larger x-ray telescopes, they will be able to take spectra of even more highly ionized oxygen and other gas.

COS will have other critical work to do besides taking the measure of the intergalactic medium and the cosmic web. For example, numerous quasar sight lines selected by astronomers will intentionally traverse the gaseous halos of distant but known galaxies. This will provide a direct indicator of the steady production of heavy elements through cosmic time, as generation after generation of stars are born and die. Astronomers can then observe the processes involved in galaxy assembly, and the feedback mechanisms that transfer gas in both directions between galaxies and the intergalactic medium.

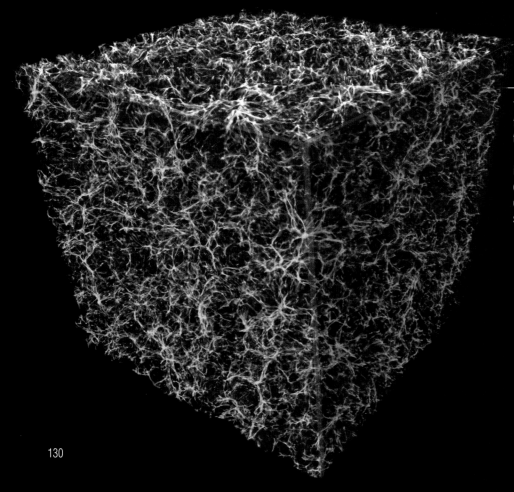

This is simulation of a slice of the cosmic web. The filaments are made mostly of dark matter located in the space between galaxies. *Hubble* and *Far Ultraviolet Spectroscopic Explorer* probed the structure of intergalactic space to look for missing ordinary matter, called baryons, that is gravitationally attracted to the cosmic web. (Figure credit: E. Hallman et al., University of Colorado.)

Further Reading

Danforth, C.W., and M.J. Shull, "Low-z Intergalactic Medium. III. H I and Metal Absorbers at z < 0.4," *Astrophysical Journal*, **679**, 194–219, 2008.

Nicastro, F., et al. "Missing Baryons and the Warm-Hot Intergalactic Medium," *Science*, **319**(5859), 55–57, 2008.

Silk, J., *On the Shores of the Unknown: A Short History of the Universe*, New York: Cambridge University Press, 2005.

Charles Danforth is a research scientist at the University of Colorado at Boulder. He grew up in northern New Hampshire, attended Swarthmore College, and received his Ph.D. from The Johns Hopkins University in 2003. At Johns Hopkins, he was active in using space-based ultraviolet instruments to study the interstellar matter, both in supernova remnants in the Milky Way galaxy, as well as larger, more energetic structures in the nearby Magellanic Clouds. After grad school, he applied his expertise with far-UV spectral data to the local intergalactic medium using both FUSE and *Hubble* observations of distant quasars. He looks forward to continuing this research with new, state-of-the-art data from the Cosmic Origins Spectrograph, a new instrument to be installed during the *Hubble* Servicing Mission 4 in 2009. His other interests include galactic structure, supernova remnants, cosmology, and hot stars. In his spare time, Charles enjoys long-distance running, mountaineering, backcountry skiing, and rock climbing.

Dr. Michael Shull is a professor of astrophysics at the University of Colorado at Boulder. He was raised in Fargo, North Dakota and St. Louis, Missouri. Dr. Shull received his B.S. in physics from Caltech, and his Ph.D. in physics from Princeton University. His theoretical and astronomy interests include studies of gas between the stars and galaxies, galaxy formation, the first stars, supernovas, and quasars. Shull is a co-investigator on the science team for the Cosmic Origins Spectrograph (COS), a new ultraviolet instrument for the *Hubble Space Telescope*. COS will provide a tenfold increase in power for studies of the distant intergalactic medium and the evolution and spatial distribution of missing matter and heavy elements.

The giant nebula NGC 3603 contains one of the most massive young star clusters in our Milky Way galaxy. It is located in one of the galaxy's spiral arms about 20,000 light-years away.

Supporting Hubble

Kelvin Garcia

Wide Field Camera 3 Integration and Test Manager
NASA Goddard Space Flight Center

Kelvin Garcia was always a tinkerer. "Growing up in Silver Spring, Maryland, I was incessantly curious about how things worked or how complex things were done. I disassembled mechanical and electrical assemblies regularly to try to figure them out. My father is an automobile mechanic. He taught me everything I know about cars and instilled in me a curiosity for how things work."

Like many other children growing up in the 1970s, Kelvin was fascinated by the space program. "In witnessing the major events in space science and exploration—the successes and failures—I decided that I didn't want to be a spectator, but instead, I wanted to be involved. The lunar landing really inspired me to become part of the kind of grand program that advances society and pushes new frontiers. The *Challenger* accident also made a deep impression on me. I realized that those astronauts who didn't make it home had risked their lives to do what they loved for the advancement of humankind."

In the end, Kelvin credits his father for his choice of careers. He worked at his dad's shop from the age of 12 to 22. "He always told me to make sure that I love what I do and that I'd go places." This led Kelvin to pursue a degree in mechanical engineering at the University of Maryland at College Park. From there, he came to work at Goddard.

In his most recent assignment, Kelvin served the *Hubble* project as the integration and test manager for the telescope's Wide Field Camera 3. In that role, Kelvin was responsible for the building and testing of this powerful, new camera. "I am honored to be a part of that team, knowing that my leadership helped to build what we all hope will be the best scientific instrument yet for *Hubble*."

Outside of work, Kelvin enjoys hiking and biking. He also strives to be the best husband and dad he can be to his wife, Tina, and their two children, Kathryn (Katie) and Josino (Jojo). "I involve myself in their school and especially their play. What better excuse can a grown-up have for playing than 'entertaining' the kids?"

Debbi Haynes-Jacintho

System Test Lead/Product Management Deputy for the *Hubble* Ground System
Lockheed Martin Mission Systems

Debbi Haynes-Jacintho always works crossword puzzles in pen and can remember every phone number she's had since she was four years old. So it was natural that the Navy, which she joined right out of high school in Lake Charles, Louisiana, assigned her to serve overseas as a cryptographic analyst. Stationed in Scotland when the Navy first started computerizing, she became a member of the first operational test team for the Naval Security Group there. "I love trying to break things, so it was a good fit for me."

Eventually, she was assigned to the National Security Agency at Fort Meade, Maryland. After leaving the Navy in 1982, Debbi became the first female test engineer at the Ford Aerospace facility in Hanover, Maryland. She spent 10 years in the DoD arena before beginning her NASA career in 1993. In her first *Hubble* position, she tested upgrades to the *Hubble*'s instrument engineering data repository.

In 1995, she joined the *Hubble* control center system team as a test engineer. Debbi now serves double duty as the system test lead and product management deputy for the *Hubble* ground system team. "I like the attention to detail that is required for testing and, working closely with software development folks like I do here, I get the added benefit of learning how things are meant to work."

She leads the team executing test procedures that exercise all of the control center system capabilities in an environment that is as close to real *Hubble* operations as possible. "The idea that I could be responsible for verifying the ground system software that controls an asset as important as the *Hubble Space Telescope* still blows me away on a regular basis."

Debbi credits her dad for her success. "He taught me early on that the easy path was not necessarily the best path, and that it was the responsibility of people who were capable of great things to go out and do them. He believed that I was one of those people, so I tried really hard to go out and do something that mattered. I joined the Navy because of my father—he was a career Navy man—and that single act put me on the path to where I am today."

Away from *Hubble*, Debbi lives the good life on Maryland's beautiful Eastern Shore with husband Tim, and Shih Tzu puppy, Maddie. She volunteers at her local animal shelter, and she enjoys interior design projects.

Bruce
Kamen

Joyce King

Systems Management and Engineering Manager
NASA Goddard Space Flight Center

Joyce King has seen *Hubble* as few others have—in the payload bay of the Space Shuttle *Discovery* prior to launch in April of 1990. She was afforded this rare opportunity as the engineer responsible for installing *Hubble* into the payload canister and preparing it for transfer to the launch pad. There, she worked with other engineers to mount *Hubble* in the Shuttle's payload bay for its historic launch. "Seeing *Hubble* installed into the payload bay was one of the most memorable experiences I have had on the program! It was truly an amazing sight!"

Joyce's *Hubble* career continued when in 1994, she moved to Maryland and began working at Goddard Space Flight Center. As the lead integration engineer for *Hubble*'s Servicing Mission 2, she headed the Goddard team of engineers and technicians preparing the *Hubble* replacement hardware and its associated support equipment for launch. After the successful mission, she moved to *Hubble* mission operations. She first served as a systems engineer and then became the deputy operations manager before transitioning to system management and engineering manager. During Servicing Missions 3A and 3B, she served as an anomaly response manager responsible for leading "tiger teams" to resolve any problems with *Hubble*.

Now, as *Hubble*'s Senior Systems Manager, Joyce provides system engineering leadership for all aspects of the *Hubble* Operations project, including daily operations, life extension, and servicing mission activities. During SM4, she will serve as the *Hubble* servicing mission operations manager on the planning shift, and will be responsible for all operations in the Space Telescope Operations Control Center at Goddard.

Joyce has come a long way from her tiny hometown of Shaftsbury, Vermont. She holds a B.S. in mechanical engineering from the University of New Haven and a M.S. in engineering management from the University of Central Florida. "I am very proud to be part of one of the greatest space exploration programs NASA has ever undertaken and feel lucky to regularly see all the wonderful images from the telescope."

To help build morale, Joyce organizes many social events on the *Hubble* project. Outside of work, she likes to cook and enjoys spending time with her son Taylor, 14, and daughter Madison, 9. One of her favorite pastimes is playing soccer, and she plays in both a spring and fall league. "I enjoy running after the ball and playing with my team. I also play volleyball, softball, tennis, and racquetball. I like to keep active—I'm not the kind of person to just sit around in front of a TV."

David Ford

Electrical Engineer
Honeywell Technical Solutions, Inc.

It seems only natural that Dave Ford would dream of a technical career at NASA. Growing up in Baltimore during the *Apollo* missions, Dave watched with fascination as the first humans walked on the Moon. Another strong influence during his early years was his father. "He was in the Navy and had taken some training in electronics. He used to repair TVs, and as a child, I would watch him do that. I also used to read his training material." Dave's love of electronics led him to attending Catonsville Community College. There he enrolled in the Cooperative Education program, which opened the door to work at Goddard Space Flight Center. "It was a dream come true."

After completing community college, Dave started at Goddard as a junior electronics technician. While working, he continued his studies at George Washington University, where he received a Bachelor's degree in electrical engineering. Dave joined the *Hubble* Team just prior to the telescope's launch in 1990. He is responsible for installing and maintaining the hardware used in the facilities, computer equipment, and networking infrastructure needed to run the telescope. "What keeps my job interesting is that we are continually implementing new technologies in the ground system. Things have really changed since when I first began on *Hubble*."

Dave is married to Darlene, his wife of 24 years. He is a believer in Jesus Christ and has served as an associate minister at the Genesis Bible Fellowship Church for six years. This involves preaching in the pastor's absence, providing marriage counseling, producing a newsletter, and leading several ministries. Dave loves studying the Bible and teaches an Old Testament class at the Baltimore School of the Bible.

Dave finds spirituality in the *Hubble* images. "When we view images of this vast, beautiful universe, it helps us to understand what we are. When we begin to understand what we are, we then we can begin to understand why we are, which I believe is the most important question a person needs to answer. What a privilege to be part of the *Hubble* program, which enables us to see more clearly and in greater detail the awesomeness of God's creation."

Lynn Carlson

Graduate Student
The Johns Hopkins University

Lynn Carlson is a storyteller. A graduate student in astrophysics at Johns Hopkins, she credits her Advanced Placement history teacher in high school with teaching her to see the story in events. "That's one of the things I find so fascinating about astronomy—it's like history, but on a grand scale and is full of amazing stories."

The story she's hoping to tell these days is how star clusters appear and develop. Her particular objects of study are clusters in the Magellanic Clouds, two nearby galaxies. Using *Hubble*, she is studying the nature and progression of star formation in the young stellar nurseries seen within these galaxies. *Hubble*'s deep, high-resolution images reveal the faint, newly formed stars, as well as other protostellar objects, all of which are beautiful and fascinating.

Lynn credits her parents for fostering her love of learning. "I was home-schooled for a while when I was young, and they always let me be self-directed and explore what interested me most at a given time." After taking physics in high school, Lynn knew she wanted to try her hand at astrophysics. She performed research in solar physics all through college, but considers the summer she spent at the High Altitude Observatory in Boulder, Colorado to be pivotal. "It was there I started to learn about how stars form. Just reading about that gave me a magical feeling, and I knew when I started graduate school that I would want to do research in this area."

Lynn earned Bachelor's degrees in both astrophysics and philosophy in the Honors College at Michigan State, where she also enjoyed studying Russian. Now busy at the Johns Hopkins University in Maryland, Lynn manages to find time to do other things as well. "I love to paint and to spend time with friends. I've also been working with some people at Goddard Space Flight Center on an astronomy program for Girl Scouts during each of the last four years and helping to form a graduate student mentoring program here at Johns Hopkins."

Lynn occasionally speaks about astronomy at Baltimore-area public middle schools, and she screens candidates for the academic Telluride Association Summer Program, something in which she participated in high school. This program creates and fosters educational communities that teach leadership and service through democratic participation. "I like to see people realize just how smart they are and that they can be leaders."

Ann Feild

Senior Graphic Designer/Illustrator for News
Association of Universities for Research in Astronomy, Inc. (AURA)

Ann Feild has come a long way since the days when she drew the Orioles' mascot for *The Baltimore Sun*. "My career's been a funny, winding road, but I've always drawn—I'm an illustrator first and foremost." These days, she employs her artistic talents as a science illustrator at the Space Telescope Science Institute in Baltimore. Ann creates graphics and illustrations for *Hubble* news releases and also designs artwork for various *Hubble* outreach efforts.

She began her art career as a medical book designer where she learned about document production and editing. Eventually, she found herself at *The Baltimore Sun*, where she spent 12 years. "I had the good fortune to do quite a few feature illustrations about lifestyle, personalities, and celebrities, and then also op-ed pieces, which I loved. And I also drew the Oriole bird. They used it in the index box of the *Sun* as a barometer of whether or not the Orioles baseball team had won or lost—so you could look at a glance and see if the little bird was happy or upset." Some of this work is on display at the Babe Ruth Museum in Baltimore.

Ann credits her family for her early introduction to space and astronomy. "My father and my brothers were very interested in astronomy, and we would go out—not just on summer evenings, but also in the winter—to look at meteor showers. My uncle by marriage, Albert Sehlstedt, reported on the NASA program for *The Baltimore Sun* in the 1960s and '70s. He covered the initial planning and development stages of *Hubble*. We followed his career with great interest. I also remember being transfixed as an adolescent by the *Apollo* missions with astronauts walking on the Moon."

Ann enjoys cooking and classical piano—something she studied for 10 years. She loves to travel and has had the opportunity to visit many countries, including Japan, Iceland, Guatemala, India, and various places in Europe. In Guatemala, she worked on a coffee plantation, as well as in an infirmary and orphanage.

In a painting class taught by her brother Christopher, Ann met an astronomer who told her about an opening at the Space Telescope Science Institute's Office of Public Outreach. Intrigued by the opportunity to be a part of this historic program, she applied and was awarded the position. Twelve years later, she still feels very fortunate to work on *Hubble*. "There's still so much to learn. The opportunities for discovery are as boundless as the universe itself."

Bill Crabb

Servicing Mission Manager
Honeywell Technology Solutions, Inc.

One of Bill Crabb's earliest memories is of standing in his backyard in Syracuse, New York with his parents, looking through a tube made of rolled up newspapers, and watching the *Sputnik* satellite pass overhead. "I think my interest in astronomy and space science really started that early." His parents bought him an inexpensive telescope when he was about seven years old. "The *Mercury* program took place during my grade school years, and I diligently watched all the flights on TV and learned about the vehicles and crews. But for me, nothing captured my imagination more than the time spent in the backyard looking through my telescope."

Bill went on to earn a college degree in astronomy and began his career at NASA in 1978. During his 26 years with *Hubble*, he has held a number of positions. Prior to *Hubble*'s launch in 1990, he was the optical telescope assembly system engineer for the mission operations contractor. About a year before launch, he became the safing system engineer. He spent much of the first year after launch pulling *Hubble* out of safe mode as the new observatory went through its growing pains.

About a year after launch, he became part of a team supporting the first servicing mission. He was assigned the responsibility for developing a plan and the products for handling contingencies that could occur during the mission. Later that year, he was promoted to deputy mission operations manager. Eventually, Bill became the contractor's servicing mission manager. In this role, he is responsible for ensuring that the required servicing mission products are developed and that the mission operations contractor team is ready to support the mission.

Outside of work, Bill participates in golf and bowling leagues, and has been an avid golfer for 40 years. He is also a woodworker who enjoys turning wood on a lathe. "I make furniture and wooden bowls, and I have a small business selling pens made from rare and exotic woods from all over the world." Bill met Susan, his wife of 29 years, during his first job at Goddard, on the *International Ultraviolet Explorer* project. They recently started working together again when Susan took a *Hubble* position in the database office at the Space Telescope Science Institute in Baltimore. The couple has two daughters, both of whom are talented singers and music teachers.

Jessica Regalado

Data Management Subsystem Engineer Lead
Honeywell Technology Solutions, Inc.

Born in Songtan, South Korea, Jessica Regalado immigrated to the United States at the age of 13. "At first, I could not get adjusted to a new life in the United States. It was very different: the language, people, house, environment—just about everything. I remember my sister and I cried every night because we wanted to go back to Korea. My mother, who was a single mom, told us that we were here so we could have an opportunity to make something of ourselves."

And given the opportunity, that is what she did. As the data-management subsystem engineer lead, Jessica now heads the engineering group that provides expertise on *Hubble*'s main computer system for all phases of telescope operation—routine, contingency, and servicing mission-related. "We are experts in onboard computer and data-recorder operations, command processing and routing, and the command/telemetry database. We also serve as test conductors for major flight software modifications and engineering tests."

In high school, Jessica's guidance counselor recognized her talent in math and science and recommended that she pursue a degree in electrical engineering. The summer after high school, Jessica was chosen for an engineering study program that included a visit to Goddard. "I remember walking through the *Hubble Space Telescope* control center and I was really excited about it! My heart rate rose and I had butterflies in my stomach. I believe that it was destiny for me to work on *Hubble*. Needless to say, I jumped at the offer to work on the program."

Jessica earned her B.S. in electrical engineering from Morgan State University and came to work on *Hubble* 13 years ago. One of her most exciting memories is from Servicing Mission 3A, during the replacement of *Hubble*'s original main computer by the astronauts. "I was four-and-a-half-months pregnant. I was nervous and excited at the same time when we had to power on the new advanced computer. As soon as that power-on command was sent, my baby kicked me hard—it was first and only time that I felt his kick. Today my son, who is eight years old, is proud of me for working on the *Hubble* Project. He knows most people recognize the *Hubble Space Telescope* and are fascinated by its discoveries."

Jessica lives in Ellicott City, Maryland, where she loves spending time with her husband George, son Leo, and mom Lisa.

Reneé Robinson

Chief Safety and Mission Assurance Officer
NASA Goddard Space Flight Center

As a young girl, Reneé Robinson was inspired by her mother, an engineer who shared her experiences with her children. "It was interesting to hear the stories she would tell about her challenges at work. It gave me something to look forward to, knowing that there would be challenges, and that there are ways of handling and dealing with them. She definitely inspired me to go out and do my best."

Reneé's passion for science and mathematics in high school, along with the example of her mother's career, motivated her to pursue a degree in electrical engineering. She graduated from the Stevens Institute of Technology in Hoboken, New Jersey. In addition to her mother, Reneé credits her tutors and mentors in the Stevens Technical Enrichment Program, as well as the National Society of Black Engineers, for helping her hone the skills that she uses today.

Born and raised in the Washington, D.C. metropolitan area, Reneé knew she wanted to return to the region after graduating from college. She dreamed of working for NASA, her first choice for employment. That dream came true as Reneé began her career with Goddard as a contractor. Eleven years later, she became a NASA civil servant. Reneé loves her work, and still considers NASA her dream job. "If you can find a job that you love, then you are indeed blessed!"

As chief safety and mission assurance officer for the *Hubble Space Telescope* Program, Reneé keeps both *Hubble* and the astronauts who service it safe. She provides the overall independent assurance that all *Hubble* hardware—flight and spare—is built and tested following the proper quality and safety protocols. "Safety is first! Assuring safety for this important project is very rewarding."

Reneé is driven by "the desire to do my best, to be successful in my career, and to make a difference." She is dedicated to motivating young people to pursue careers in science and engineering. "I feel like the work we are doing today is for the youth of tomorrow. I hope to be a part of inspiring young people to go after their dreams."

Away from Goddard, Reneé works with the youth at her church. She assists in coordinating the daily activities for the younger children and also works with the teenagers. Reneé is a single mom to three wonderful children—Marcus, Jasmine, and Victoria.

Nikhil Padmanabhan

Hubble Fellow
Lawrence Berkeley National Laboratory

Nikhil Padmanabhan was in middle school when he read two books by the legendary physicist Richard Feynman. "It was then I first realized that physics was something people did as a job, and that it was fun. And I've been hooked ever since."

Nikhil is now a Hubble Fellow at the Lawrence Berkeley National Laboratory. Hubble Fellows are a group of about 10 postdoctoral researchers competitively selected each year by the Space Telescope Science Institute and funded by NASA grants for three years. The Hubble Fellowships allow postdocs to pursue their own research interests—that is, not tied to a particular project or person—at an institution that best complements their research. "This approach, in my opinion, is one of the greatest strengths of the program in that it encourages fundamental astronomical research. While this may not have an immediate effect on current NASA programs, it is important for successfully planning the next generation of experiments and surveys."

Nikhil is an observational cosmologist. "My particular interest is in taking observations that help prove or disprove various theories that predict the matter and energy content of the universe and its evolution—questions like what fraction of the universe is made of atomic stuff we basically understand, and what part is not; how did galaxies form and evolve; and what is the final fate of the universe? One of the biggest endeavors in cosmology—and a large part of my research program—is to better measure the increasingly evident accelerating expansion rate of the universe. This should shed much-needed light on the complicated physics underlying it."

Born in Bombay (now Mumbai), India, Nikhil spent most of his life in Delhi before entering college at Stanford. "I came to Stanford as a undergraduate determined to try my hand at doing research, but I had little idea of what it entailed or what the various options were. So I asked my freshman physics professor to suggest a few faculty members he thought would be good people to do research with. The first person on that list was a cosmologist. I walked into his office and thought what he was doing was interesting. So I started working with him, found I enjoyed it, and the rest sort of then fell into place." After earning his B.S. at Stanford, he received his Ph.D. at Princeton before joining Lawrence Berkeley Laboratory.

Outside of work, Nikhil can often be found out hiking or reading in a neighborhood coffee shop. "I'll read just about anything, but my current two weaknesses are historical fiction and U.S. history."

Antonella Nota

Hubble Space Telescope Mission Manager for the European Space Agency
European Space Agency

Since middle school in Venice, Italy, astronomy has been Dr. Antonella Nota's passion. As a teenager, she was part of an amateur astronomy group that observed variable stars from the shores of the Venice Lagoon. "We monitored variable stars that are visible by eye or with small binoculars or a small telescope. I was the second female amateur astronomer to observe with this group at the Lagoon, after the daughter of [esteemed British astronomer] Fred Hoyle. Astronomy has always been the love of my life."

After graduating from the University of Padua, Antonella came to the Space Telescope Science Institute (STScI) as a European Space Agency (ESA) post-doc working on *Hubble*'s Faint Object Camera. "I arrived in 1986, and the *Challenger* disaster happened almost immediately afterward, so everything was put on hold. *Hubble* was finally launched in 1990. So I've been here since the beginning of its science mission."

Antonella soon acquired more and more responsibility in instrument managerial roles. Now, as the *Hubble Space Telescope* mission manager for the European Space Agency, she is a member of STScI senior management. In this role, she has oversight of the ESA operations at STScI. "ESA is a 15-percent partner with NASA on the *Hubble* program. They contributed with hardware, money, and now support personnel. Today, with 15 scientists here, ESA scientists comprise 30 percent of the total number of scientists working on *Hubble* at STScI." Antonella supervises these scientists, and she also manages all of the issues concerning the international relationship with ESA for *Hubble*.

Star formation in stellar populations is Antonella's specific area of astronomical interest. She is particularly fascinated by an image of the Small Magellanic Cloud—a nearby galaxy—resulting from one of her own *Hubble* observations. There, significant star formation is occurring where it wasn't expected. "It's a fascinating and beautiful place. What made it explode with stars and why?"

Antonella enjoys being an advocate for women in science. "I give speeches, mentor, and work on programs that could expand the pool of women in science. I also try to address some of the issues that create obstacles for women." Antonella has a lifelong love of art and classical music. She's also been known to dabble in extreme sports, including scuba diving and skydiving.

She is married to another astronomer, Dr. Mark Clampin, who is the observatory scientist for the *James Webb Space Telescope* at Goddard. They have an eight-year-old daughter, Simona.

Sharon Dixon

Senior Project Engineer, *Hubble* Program
Ball Aerospace & Technologies Corporation

Growing up in Chicago, Sharon Dixon wanted to be a teacher, but couldn't shake her intense desire to pursue math and science. "I was always attracted to the essence of engineering—exact science and reproducible results." Following high school graduation, Sharon enrolled at an Arizona community college to study electrical and electronic technology and found herself literally making bread to make bread—supporting herself by working the night shift in the bakery of a local grocery store. "I was working full time and carrying a full load of classes and always tired. One morning, that resulted in donuts sprinkled with salt, rather than sugar. It's funny now, but that morning it felt pretty career limiting."

Her Associate's degree landed her a job at Sperry Flight Systems, and she was soon enrolled at Arizona State University where she earned her Bachelor's degree in electrical engineering. Sharon joined Ball Aerospace in 1986, and shortly thereafter, began working the *Hubble* program. As fate would have it, Sharon's early teaching instincts have been put to good use. When Servicing Mission 4 was reinstated, she was asked to mentor and coach others at Goddard. She was soon affectionately known as "Miss Ball" for her knowledge of Ball's *Hubble* instrument electronics and system integration, as well as her ability to teach others.

As a senior project engineer for Ball Aerospace, Sharon served as the electrical lead for three *Hubble* instruments: the Cosmic Origins Spectrograph, the Wide Field Camera 3, and the Space Telescope Imaging Spectrograph. Since 2000, she has been responsible for testing more than 100 non-flight and flight printed wiring assemblies for *Hubble*. Sharon's hard work earned her a Goddard superior performance award. At Goddard or Ball, Sharon's approachability and experience make her a favorite of junior technicians, assemblers, and inspectors who seek her out.

Ten years ago, Sharon's devotion to inspiring future generations led her to organize the first "Take Your Daughters to Work Day" at Ball, creating an annual event that now draws more than 300 boys and girls. She is equally active in her own children's lives and proud that all three are pursuing engineering degrees.

Away from work, Sharon says hiking, biking, and camping compete for her free time with home improvement projects, landscaping, and gardening. She is married to a Ball engineer and cheerfully admits that she sometimes feels *Hubble* consumes her life. "All my coworkers are my best friends!"

Dorothy Fraquelli

Archive Scientist
Computer Sciences Corporation

Dr. Dorothy Fraquelli has been enthralled by astronomy for as long as she can remember. "As a child, I would go out and watch lunar eclipses every chance I could." Although she earned a Ph.D. in astronomy from the University of Toronto, she never envisioned herself working on a space telescope. But she came to the Space Telescope Science Institute (STScI) in October of 1984—six years before the telescope launched—and has been with *Hubble* ever since.

Dorothy began her *Hubble* career by testing the observational support system of the original science data-processing ground system, the widely diverse collection of facilities, computer equipment, and networking infrastructure needed to process data from the telescope. In her current role, she helps outside users retrieve data from the archive, which contains all of *Hubble*'s data. She is responsible for improving the archive and finding ways to make the data more useful. Dorothy is the liaison between archive operations and the developers. She also runs the coordination meeting among STScI and its international partners: the Space Telescope European Coordinating Facility, and the Canadian Astronomy Data Centre.

She credits her Catholic grade school education in Royal Oak, Michigan with her strong background in science and math. "Just about all the teachers were nuns. They taught math, they taught science, and they expected you to learn it. There wasn't any of this 'I can't do that because I'm a girl.' I remember the nuns bringing television sets into the classrooms so we could watch the launches. These were the very early missions, the *Mercury* and *Gemini* missions."

One of Dorothy's fondest *Hubble* memories is of all of the excitement surrounding the impact of comet Shoemaker-Levy with the planet Jupiter in 1994. Dorothy was running the shift in the Institute's operations area, which receives *Hubble*'s science and engineering data. "The whole science team filled the operations area. When the first impact images were displayed, the team was ecstatic, discussing what they were seeing, what it meant, and predicting what the next images would show."

In her spare time, Dorothy enjoys theater and is an avid reader. She considers her reading choices "eclectic," including science fiction, history, and biography, and she even dabbles a bit in the biological sciences. As a board member of the Friends of the Towson Library, she works on projects and programs to improve the library. For more than 25 years, she has also belonged to the Baltimore Branch of the American Association of University Women, which supports education and equity for all women and girls.

Known as the Antennae galaxies, NGC objects 4038/4039 are two spiral galaxies in the process of colliding. Their two yellowish cores, seen crisscrossed by brown filaments of dust surrounded by pink-glowing hydrogen gas, consist mainly of old stars. Many young, blue star clusters are also evident. The Antennae galaxies are approximately 62 million light-years from Earth, located in the constellation Corvus.

Acknowledgments

Acknowledgments

Credit for the success of the *Hubble Space Telescope* rightly belongs to an entire universe of people and organizations. First and foremost are the citizens of the United States and Europe, who have steadfastly supported *Hubble* over the years with their tax dollars and their enthusiasm. As a result, thousands of astronomers from around the world have successfully used *Hubble* to probe the deepest mysteries of the universe and have shared their discoveries through both professional publications and public outreach. Educators and students worldwide have recognized in *Hubble* an important source of knowledge, excitement, and motivation about science.

A small cadre of astronauts from NASA and the European Space Agency (ESA) have taken significant personal risk to service *Hubble*, maintaining and upgrading the spacecraft to keep it at the forefront of astronomical research. The Science Mission Directorate at NASA Headquarters, and the HST Program Office at NASA's Goddard Space Flight Center have led the *Hubble* program over the years, with major contributions to the observatory—both hardware and people—also provided by the ESA.

Hubble's highly successful science program has been organized and guided by the Space Telescope Science Institute, operated by the Association of Universities for Research in Astronomy, under contract to NASA. Last, but not least, many dedicated NASA employees and dozens of first-class contractor organizations throughout the global aerospace industry have designed, built, and successfully operated *Hubble* and its scientific instruments over a period spanning decades.

All these people and organizations should take pride in the scientific achievements described in this publication.

For additional information, contact:

Susan Hendrix
NASA Goddard Space Flight Center
Office of Public Affairs
Greenbelt, MD 20771
301-286-7745

Space Telescope Science Institute
3700 San Martin Drive
Baltimore, MD 21218-2410
410-338-4444 (general info)
410-338-4707 (technical info)

http://hubble.nasa.gov/
http://hubblesite.org/

The team at Space Telescope Science Institute for this publication included Ann Jenkins (Writer), Ann Feild, Mario Livio, Sharon Toolan, Ray Villard, Jill Lagerstrom, and Donna Weaver. The team at Goddard Space Flight Center included Kevin Hartnett (Lead), James Jeletic, David Leckrone, Malcolm Niedner, Mindy Deyarmin, Michael Marosy, Pat Izzo, Elaine Firestone, Sue Goldberg, and Sherri Panciera.

In reference lists, please cite this document as:

NASA, *Hubble 2008: Science Year in Review*, K. Hartnett (ed.), NASA Pub. 2009-1-069-GSFC, NASA Goddard Space Flight Center, Greenbelt, Maryland, 152 pp., 2009.